Timo Busch

Carbon Constraints

Timo Busch

Carbon Constraints

Corporate Strategies, Risk Management, and Firm Performance

Südwestdeutscher Verlag für Hochschulschriften

Impressum / Imprint
Bibliografische Information der Deutschen Nationalbibliothek: Die Deutsche Nationalbibliothek verzeichnet diese Publikation in der Deutschen Nationalbibliografie; detaillierte bibliografische Daten sind im Internet über http://dnb.d-nb.de abrufbar.
Alle in diesem Buch genannten Marken und Produktnamen unterliegen warenzeichen-, marken- oder patentrechtlichem Schutz bzw. sind Warenzeichen oder eingetragene Warenzeichen der jeweiligen Inhaber. Die Wiedergabe von Marken, Produktnamen, Gebrauchsnamen, Handelsnamen, Warenbezeichnungen u.s.w. in diesem Werk berechtigt auch ohne besondere Kennzeichnung nicht zu der Annahme, dass solche Namen im Sinne der Warenzeichen- und Markenschutzgesetzgebung als frei zu betrachten wären und daher von jedermann benutzt werden dürften.

Bibliographic information published by the Deutsche Nationalbibliothek: The Deutsche Nationalbibliothek lists this publication in the Deutsche Nationalbibliografie; detailed bibliographic data are available in the Internet at http://dnb.d-nb.de.
Any brand names and product names mentioned in this book are subject to trademark, brand or patent protection and are trademarks or registered trademarks of their respective holders. The use of brand names, product names, common names, trade names, product descriptions etc. even without a particular marking in this works is in no way to be construed to mean that such names may be regarded as unrestricted in respect of trademark and brand protection legislation and could thus be used by anyone.

Verlag / Publisher:
Südwestdeutscher Verlag für Hochschulschriften
ist ein Imprint der / is a trademark of
AV Akademikerverlag GmbH & Co. KG
Heinrich-Böcking-Str. 6-8, 66121 Saarbrücken, Deutschland / Germany
Email: info@svh-verlag.de

Herstellung: siehe letzte Seite /
Printed at: see last page
ISBN: 978-3-8381-0677-9

Zugl. / Approved by: Zurich, ETH, Diss., 2008

Copyright © 2009 AV Akademikerverlag GmbH & Co. KG
Alle Rechte vorbehalten. / All rights reserved. Saarbrücken 2009

Acknowledgements

Driven by the increasing salience of climate change as one of the chief global challenges facing society in the 21st century, I have focused since my early research on the concept of carbon constraints as a new issue that needs to be addressed by firms' strategic management. From the beginning of my doctoral studies at ETH Zurich in 2005, I found an excellent academic environment in which to develop this concept and write several peer-reviewed research articles in this field. Now, in 2008, I look back at a challenging, yet highly rewarding endeavor that would not have been possible without the enduring support of several individuals.

I would like thank my supervisor Prof. Volker H. Hoffmann for giving me the opportunity to do this kind of research – in general – and supporting my research ideas – in particular – with outstanding feedback and literally never-ending patience. Also, I would like to express my gratitude to the entire Sustainability & Technology chair at ETH Zurich. Each individual colleague made his contribution to this dissertation, though on different levels, but with tangible impact. Furthermore, I am grateful to Prof. Roland W. Scholz for his very helpful comments and recommendations that significantly helped me to finish the dissertation.

Through presentations at several international research conferences and in the process of organizing the ETH PhD-Academy on Sustainability & Technology, I was in touch with an enormous number of scholars around the globe. It would be a never-ending task to single out individuals at this point. However, I would like to acknowledge their manifold and excellent feedback, inspiration, and discussions, which definitely stimulated my thinking and helped me find my way through the academic jungle. Furthermore, I am particularly indebted to the Stiftung Drittes Millennium, which made a generous donation to ETH Zurich that lead to the foundation of the research group for Sustainability and Technology and, thus, made my research possible.

Finally, I owe my biggest thanks to my family, notably my parents, Nina and Julian, for their understanding of PhD-implicit time-constraints and their mental support as well as their – often crucial for survival – welcome distraction throughout my studies.

Abstract

The current debates about increasing oil prices and climate change have one thing in common: both center around carbon. Considering the former, carbon scarcity regarding fossil resources emphasizes that we do not have enough of it. Considering the latter, carbon abundance in the atmosphere illustrates that we have too much of it. In sum, the main issue is that the accumulation of carbon is occurring at the wrong place. Therefore, the dependency on carbon-based materials and energy sources as well as the significant reduction of greenhouse gas emissions have been recognized as major challenges of the 21st century. This dissertation sheds light on these phenomena from an industrial ecology perspective and investigates the consequences for strategic management.

Firms are central to the effort to grapple with these carbon challenges due to the large material flows they process and their capabilities for technological innovation. But most importantly, the carbon developments primary taking place within the natural environment alter the business environment: fossil fuel scarcity results in significant price increases on energy markets; climate change has direct and indirect effects that change established production and consumption patterns. As such, firms face the emergence of carbon constraints that need to be addressed by corporate strategy. By effective carbon management and substantial efforts to curb greenhouse gas emissions, firms are therefore not only central in contributing to a significant decarbonization of society; they also pave their own way for future profitability and competitiveness. However, along with this are several implications for strategic management that are addressed by my research.

First, carbon constraints constitute an emerging business issue and thus need to be assessed. In this dissertation, four carbon performance indicators are developed in order to provide strategic management with the data necessary for carbon-related assessments and decisions. The indicators illustrate how undertaken measures result in improvements in the firm's carbon performance. Furthermore, it is elaborated on how policy makers as well as investors and financial institutions can use such information within their strategic decision processes.

Second, carbon constraints shape the business environment in areas that previously have been taken for granted and are furthermore subject to uncertainty. Thus, the consequences for strategic management are discussed from a corporate risk perspective. Notably, an investment framework is developed in order to facilitate the incorporation of ecology-induced uncertainties such as those that emerge in the context of carbon constraints.

Zusammenfassung

Die derzeitigen Debatten über steigende Ölpreise und den Klimawandel haben folgende Gemeinsamkeit: Beide stehen in Beziehung zu Kohlenstoff. Einerseits veranschaulicht die Knappheit fossiler Rohstoffe, dass nicht genug Kohlenstoff-Reserven vorhanden sind, andererseits verdeutlicht der Klimawandel, dass sich zu viel Kohlenstoff in der Atmosphäre befindet. Das Hauptproblem besteht darin, dass sich die Akkumulation von Kohlenstoff an der falschen Stelle vollzieht. Die Abhängigkeit von Kohlenstoff-basierten Materialien und Energiequellen sowie das Emittieren von Treibhausgasen werden daher als die zentralen Herausforderungen des 21. Jahrhunderts angesehen. Diese Dissertation betrachtet diese Phänomene aus einer Industrie-ökologischen Perspektive und untersucht die sich daraus ergebenen Konsequenzen für das strategische Management.

Bei der Lösung dieser Kohlenstoff-bezogenen Herausforderungen nehmen Unternehmen eine zentrale Rolle ein, da sie für einen Grossteil der Materialflüsse verantwortlich sind und sie die entsprechenden Fähigkeiten besitzen, um die notwendigen technologischen Innovationen umzusetzen. Am wichtigsten ist jedoch, dass diese, primär in der natürlichen Umwelt stattfindenden, Kohlenstoff-bezogenen Entwicklungen auch das betriebliche Umfeld verändern: Eine Verknappung fossiler Rohstoffe geht einher mit signifikanten Preissteigerungen auf den Energiemärkten. Ebenso hat der Klimawandel direkte und indirekte Effekte auf etablierte Produktions- und Konsummuster. Daher ergeben sich für Unternehmen neue, Kohlenstoff-bezogene Beeinträchtigungen, die innerhalb von strategischen Entscheidungen berücksichtigt werden müssen. Durch ein effektives Kohlenstoffmanagement und substantielle Anstrengungen, die Emission von Treibhausgasen zu reduzieren, spielen Unternehmen nicht nur eine zentrale Rolle für eine Reduzierung der Kohlenstoff-Abhängigkeit der Gesellschaft, vielmehr verfolgen sie so auch gleichzeitig ihre eigenen Ziele hinsichtlich zukünftiger Profitabilität und Wettbewerbsfähigkeit. In diesem Kontext ergeben sich allerdings einige Management-bezogene Implikationen, die im Rahmen meiner Dissertation diskutiert werden.

Erstens werden Indikatoren als Bewertungsgrundlage für strategische Managemententscheidungen benötigt, die Aufschluss über Kohlenstoff-bezogene Entwicklungen geben. Zu diesem Zweck werden vier Indikatoren entwickelt, die aufzeigen, wie effektiv einzelne betriebliche Massnahmen zu einer Verbesserung der Kohlenstoffflüsse führen. Darüber hinaus wird diskutiert, wie politische Entscheidungsträger sowie Investoren und Finanzdienstleister derartige Informationen innerhalb ihrer eigenen strategischen Entscheidungsprozesse benutzen können.

Zweitens werden neue betriebliche Instrumente benötigt, um Kohlenstoff-bedingte Veränderungen adäquat zu managen, da diese Veränderungen Bereiche betreffen, die zuvor keine besondere Berücksichtigung innerhalb des strategischen Managements bedurften und zudem ihre zukünftige Entwicklung schwer abzusehen ist. So ergeben sich neue Unsicherheiten, deren Konsequenzen hinsichtlich des betrieblichen Risikomanagements diskutiert werden. Insbesondere wird ein Handlungskonzept für Investitionsentscheidungen entwickelt, das die Berücksichtigung von ökologisch bedingten Unsicherheiten, wie sie zum Beispiel im Kontext von Kohlenstoff-bedingten Veränderungen entstehen, ermöglicht.

Table of Contents

Acknowledgements i
Abstract ii
Zusammenfassung iii
Table of Contents v
1　Introduction 1
2　Objectives 4
3　Methods 8
4　Summaries of the Papers 9
5　Conclusion 11
6　Papers 19
Endnotes 20
Annex

1 Introduction

The phenomena of high crude oil and gas prices and the prevailing public and scientific debate about global warming have one thing in common: both center around carbon. Considering the former, carbon scarcity regarding fossil resources emphasizes that we do not have enough of it. Considering the latter, carbon abundance in the atmosphere illustrates that we have too much of it. In sum, the main issue is that the accumulation of carbon is occurring at the wrong place. The carbon was accumulated in the ground in great quantity over geologic time-scales, and now this quantity it is being transferred to the atmosphere in a, comparatively, overwhelming short time period. Therefore, overcoming the dependency on carbon-based materials and energy sources as well as significantly reducing greenhouse gas emissions have been recognized as major challenges of the 21st century.[1] This dissertation considers the firm-level of analysis in the context of the emerging carbon constraints.

The four main words in the dissertation's title portray the general frame of the research: carbon, constraints, strategic, and management. By *carbon* I refer to an input and an output dimension, i.e. a firm is considered as an organizational entity that requires carbon-based inputs and releases carbon-containing outputs. The input dimension relates to production processes that utilize carbon-based materials and energy. Here, fossil fuels and fossil-fuel based inputs such as crude oil-based plastics or coal-based electricity are considered. The output dimension refers to the emission of greenhouse gases in these production processes, the most important part of which relates to CO_2 emissions.[2]

While discussions about global sustainability challenges abound, the financial effects that the underlying developments and restrictions incur, albeit important, have received less attention. In this context, within this dissertation *constraints* are defined as any sort of condition that limits firms' ways of conducting business and their efforts towards profitability. With regard to carbon, on the input side carbon constraints are related to the disposition of fossil fuels. These are determined and influenced twofold, by price increases stemming from natural scarcity as well as by socio-political factors such as changing consumption patterns towards low-carbon products. Carbon constraints on the output side can be divided into direct and indirect climate change effects. The former describe direct physical impacts of climate change on a firm's assets and processes and the latter encompass impacts due to human efforts such as climate policy to mitigate global warming.

By emphasizing the *strategic* dimension of my research, the title denotes the nature of the research theme. Although carbon resource extraction and climate change have to be considered as rather short-term developments in geological terms, both phenomena become relevant rather in a the long-term from a society or business perspective. Global warming takes place over decades, and the effects of current emissions in terms of their societal and economic consequences will be most tangible for future generations. The same holds for natural resource endowments. Although there is no doubt that all natural resources are limited, it can be a rather long-term process – determined mainly by rate of growth in resource consumption and technological trajectories – until a single reserve is finally depleted. Consequently, I base my arguments on current developments and put them in a forward-looking perspective. I make the claim that while the adverse effects of climate change and fossil fuel depletion are known in a long-term, i.e., strategic perspective, the vision of a decarbonizing the society is still disconnected from the more tactical and operational level. As a conclusion, I argue that comprehensive action in terms of fostering low-carbon efforts is needed now.

Firms are central to the effort to grapple with these issues due to the large material flows they process and their capabilities for technological innovation. Therefore, this dissertation puts special emphasis on *management* efforts to address and incorporate carbon constraints: on the one hand, corporate management can also be seen as responsible in terms of corporations' fiduciary duty to shareholders to foster low-carbon corporate strategies since these are becoming relevant for the bottom line, notably in carbon-intensive industries. Furthermore, it can be argued that corporate management has the societal responsibility to ameliorate the negative external effects firms are causing. On the other hand, policy makers as well as investors and financial institutions can be considered as stakeholders who wield power over firms and therefore put them in a position of great dependency.[3] As such, these stakeholders' understanding and management of carbon constraints determine the pace of firms' implementation of carbon management strategies and the nature of their response to climate change issues.

In its general approach towards the role and function of firms, my research focuses on the magnitude of an emerging business issue that stems from changes in the natural environment and managerial responses to this issue. In the literature on sustainable development and corporate social responsibility, much work has been done on what firms *should* do in this regard from a normative perspective. In most of the concluding remarks of the individual papers in this thesis, however, I illustrate what the results imply for corporate risk management and from a financial markets' perspective. To this end, I consider corporate carbon management and efforts to curb greenhouse gas emissions from the standpoint of firms' profitability and their potential for cost reductions. Hence, I have chosen an *instrumental*[4] perspective on the linkage between business and the natural carbon world.

This is a cumulative dissertation, i.e. I have authored or co-authored five journal articles in total. Chapter 2 lays out the foundation of my work by discussing my research objectives, while Chapter 3 summarizes my methodological approaches. The individual papers are summarized in Chapter 4. In Chapter 5 I illustrate the contribution of this dissertation to the research field of organizations and the natural environment, discuss the concept of carbon constraints from an industrial ecology and management theory standpoint, and end with conclusions for future research. The Chapter 6 lists the papers, which are attached in the annex as published works or submissions in the process of being published.

2 Objectives

With respect to the interplay of the four components of this research introduced above, the term *carbon constraints* reflects the implication for strategic management that firms experience a carbon-related change in the business environment. This change has three main characteristics: it pertains to previously taken for granted business conditions, how it occurs at present and in the future is subject to uncertainty, and it requires new approaches from strategic management in order to address it adequately.

First, the utilization of fossil fuels and the emission of greenhouse gases have been *business conditions* taken for granted in the past. However, a shift in these conditions is underway, induced by a boost in the financial relevance of carbon. As this shift is driven by rather long-term developments, firms have yet to experience the full importance of carbon for financial success. Nevertheless, recent developments set a clear signal: On the carbon input side, the business environment is being influenced by price increases on energy markets: a significant demand growth in crude oil takes place while supply capacities remain more or less at their current level.[5] The theory of natural resource economics predicts price increases as a consequence of a mismatch between market supply and demand.[6] Especially in recent years, the crude oil spot price has experienced a significant upward trend.[7] Since gas prices are linked to oil prices, prices for gas have also increased.[8] As one consequence, electricity prices have increased as well in many countries.[9] On the carbon output side, firms have increasingly seen stakeholders putting climate change on their agendas: for example, in 2005 the European Commission introduced the European Emission Trading Scheme, which regulates the emission of greenhouse gases.[10] Many consumers have started to consider greenhouse gas emissions as one determinant of their purchasing decisions.[11] Furthermore, financial markets have discovered climate change as an important issue.[12] In sum, carbon constraints have started to emerge as a new business-relevant issue, as they pertain to conditions of the business environment that firms could ignore in the past.

Second, the described changes are subject to *uncertainty*. On the carbon input side, this becomes most tangible in connection with natural resource endowments. The availability of all fossil resources is naturally limited in the long-run, with crude oil as one key carbon input for economies considered by many to be about to peak.[13] Many ecological economists assume that peak oil – understood as the maximum amount of oil produced in a year – will go along with serious price increases on the oil market. However, opinions differ tremendously concerning when this peak will occur or if it has even occurred already. Thus, it is uncertain whether fossil fuel and energy prices will remain more or less stable. On the carbon output side, policy makers have started to take up the carbon challenge: For example, the European Union has set challenging greenhouse gas reduction targets for the time frame up to 2020 and beyond.[14] Regarding the international climate policy process, the USA agreed to be part of a post-Kyoto treaty, which will be negotiated in Copenhagen in 2009. However, the consequences of these future climate policy processes on the business environment are hard to anticipate. Similarly, the future effects of changing consumption patterns towards low-carbon products or the carbon-related impacts of developments in financial markets are uncertain. In sum, it is uncertain when and how severely the business environment will change due to emerging carbon constraints.

Third, as the described carbon-related developments and corresponding uncertainties become increasingly relevant to businesses, they cannot be neglected by firms that rely on utilizing and/or emitting carbon. However, strategic management could neglect these issues in the past, and the phenomenon of related uncertainties was accordingly absent. Thus, firms require *new approaches* towards assessing the implications of carbon constraints for their business and deriving corresponding management strategies. The same argument can be made for policy makers as well as investors and financial institutions: as carbon was not of prime concern in the past, they also need new approaches that facilitate their understanding of how carbon constraints (can) affect firms' decision processes and their financial performance. Based on such information, they can enhance their own carbon-related decisions processes and strategies. Developing such approaches is challenging: not only should the approaches reflect carbon constraints that relate to a firm's own production processes; they also have to display carbon risks that can lurk in its value chain. In sum, the life-cycle wide material flow assessment of carbon constraints and integration of corresponding management strategies constitute a new requisite for firms internally as well as for their external stakeholders.

Although carbon constraints are rather long-term by nature and related changes appear in many areas uncertain, the examples discussed above illustrate that the business environment is already changing and that firms need to respond to this change. This requires long-term adjustment processes, as established production and consumption patterns with respect to carbon will not shift overnight. Therefore, it is urgent for strategic management to start rethinking and altering established carbon utilization strategies now. Notably, as short-term adjustments made at a later time bear the risk to result in sub-optimal solutions. This is the case, for example, when an ad hoc response to carbon issues requires an investment in a new production engine, even though the existing engine has not amortized yet. In this respect and derived from the three delineated characteristics of the carbon-related change in the business environment, the objective of this dissertation is thus to answer the following overall research question:

How can emerging carbon constraints be addressed in strategic management?

This dissertation consists of five papers that have been published or submitted to peer-reviewed journals. Each of the papers contributes to answering this research question, but each paper on its own has a very specific focus and addresses specific aspects of the three main characteristics of the carbon-related change in the business environment. Figure 1 provides an overview of the research design of the dissertation as a whole and the research questions of the individual papers treated separately.

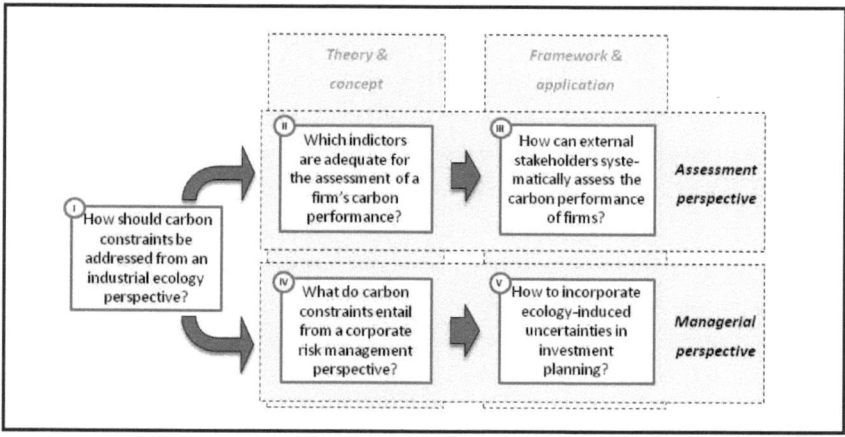

Figure 1: Research design of the dissertation and research questions of the individual papers

Paper I focuses on the implications of carbon constraints for society in general and contributes to answering the overall research question by emphasizing the relevance of managing carbon from an industrial ecology perspective.[15] Everything that is intended to be managed needs to be measured. Thus, paper II and III contribute to answering the overall research question from an assessment perspective by focusing on how to measure carbon constraints and convey the findings to strategic management. To this end, paper II derives four indicators that facilitate the assessment of a firm's carbon performance. Paper III illustrates how external stakeholders can systematically assess the carbon performance of firms by applying these indicators for US electricity producers.

In addition to this assessment perspective, papers IV and V focus on how to effectively manage the changes in the business environment due to carbon constraints. Both papers contribute to answering the overall research question from a managerial perspective by particularly investigating how the uncertainty aspect of carbon constraints can be incorporated in strategic management. Paper IV scrutinizes the emergence of carbon constraints in general and derives recommendations for corporate risk management. Paper V describes the general managerial challenge of incorporating ecology-induced uncertainties and exemplifies how to proceed appropriately in the context of carbon constraints and investment planning.

3 Methods

Throughout the five papers written under the umbrella of this dissertation, different methods with respect to the different research questions have been applied. Figure 2 illustrates the research methods of the papers. Paper I serves as a general introduction to the topic and makes a normative claim of what should be done in the context of carbon constraints and the challenge of global sustainable development.

Paper II and IV are theoretically driven and develop new conceptual approaches regarding the assessment and analysis of carbon constraints. Paper II derives definitions for performance indicators from a literature review of the broad field of carbon management, greenhouse gas assessments, and related corporate tools. Paper IV defines the areas of carbon constraints for corporate risk management based on a literature review of corporate risks relating to climate change and carbon utilization.

Paper III and V are both application-oriented and suggest frameworks as new strategic management approaches for assessing and incorporating carbon constraints. Paper III applies the previously-defined carbon performance indicators empirically by conducting a quantitative analysis of the 100 largest US electricity producers and by deriving a carbon risk ranking. Paper V investigates corporate risks stemming from the natural environment by analyzing management perceptions of ecology-induced uncertainties, which are illustrated by examples in the carbon constraints context.

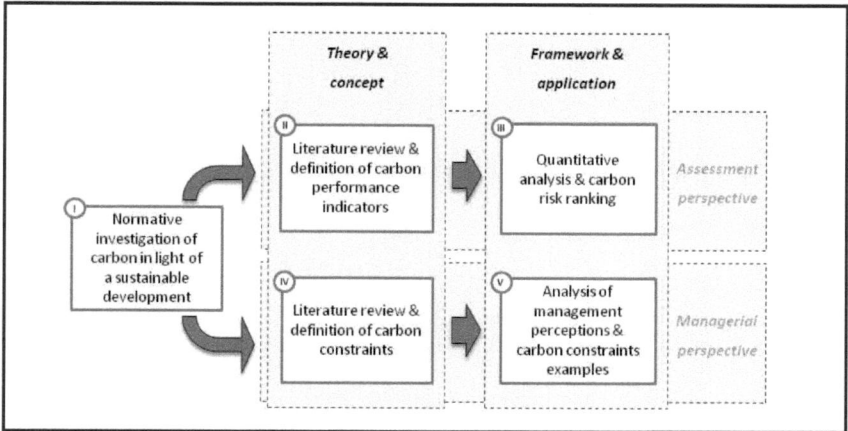

Figure 2: Research methods of the papers

4 Summaries of the Papers

Paper I derives carbon constraints as a global issue mankind is facing in the 21st century. It poses the question whether carbon-related issues such as fossil fuel scarcity or climate change are totally new. By referring to Easter Island, it considers how carbon-related issues emerged earlier in history. On this island, the low availability of carbon constrained the activities of the islanders: first, as the most visible sign of economic breakdown, the carving of statues stopped; later, the entire society collapsed. It is discussed how our modern society is on a similar track, just with a new scope and on a different scale. As a conclusion it is suggested to investigate carbon constraints and their uncertainties on different economic levels in order to prevent them from again becoming a critical issue, notably on a global scale. As such, this paper illustrates the general motivation for this dissertation and poses the foundation for its research questions.

Paper II focuses on the assessment perspective and defines four carbon performance indicators. As carbon constraints emerge, it is important to determine the individual stake firms have in these issues and to measure firms' carbon performance. The paper defines comprehensive and systematic corporate carbon performance indicators: (1) Carbon intensity is physically oriented and represents a firm's carbon use in relation to a business metric. (2) Carbon dependency illustrates the change in physical carbon performance within a given time frame. (3) Carbon exposure reveals the financial implications of using and emitting carbon. (4) Carbon risk estimates the change in financial implications of carbon usage within a given time frame. Based on these general definitions, the paper specifies the indicators for a standardized application. Corresponding assessment results provide firms information for strategic management decisions as well as support for two important external stakeholders in their strategic decision making: policy makers can include such information when evaluating current climate policies and formulating future ones; investors and financial institutions can compare firms with respect to their carbon performance and corresponding financial effects.

Paper III applies the previously defined carbon performance indicators by suggesting an assessment framework and analyzing electricity producers' carbon performance. As such, this paper takes an external perspective and delivers information for stakeholders' strategic decision making processes. While the US Energy Information Administration's (EIA) 'Annual Energy Outlook 2007 – Reference Case' is based on constant energy policy and market conditions, the US Senate bill 1766, the 'Low Carbon Economy Act of 2007,' assumes a more carbon-constrained business environment. Based on these two market forecasts and different corporate carbon strategies, carbon scenarios are derived and used as the basis for an empirical carbon performance assessment of the 100 largest US electricity producers. The assessment results discuss in detail the firms' physical carbon performance and the associated value in monetary terms. One important result of the paper is a carbon risk ranking of the electricity producers, which notably provides important information for actors on financial markets and their strategic investment decisions.

Paper IV transfers the phenomenon of carbon constraints to corporate risk management. A comprehensive carbon risk assessment takes into account input- as well as output-related factors. Thus, this paper defines carbon constraints as phenomena which emerge due to the disposition of fossil fuels in the input dimension and due to direct and indirect climate change effects in the output dimension. Based on a literature review regarding the financial consequences of carbon constraints on the macroeconomic, sector, and firm level, the following conclusions are drawn: First, financial consequences seem to be asymmetrically distributed between and within sectors. Second, the individual risk exposure of firms depends on the intensity of and dependency on carbon-based materials and energy, which is investigated in-depth in papers II and III. Third, financial markets have only started to incorporate these aspects in their valuations. Paper IV ends with recommendations on how to incorporate our results in an integrated carbon risk management framework.

Paper V also focuses on the managerial perspective and suggests real options as a management approach for incorporating ecology-induced uncertainties such as those in the context of carbon constraints. The initial argument is made that the role of uncertainty in general features prominently in the management literature, but how corporate investments should proceed in the face of uncertainties relating to the natural environment is less discussed. From the perspective of ecological economics, the salience of ecology-induced issues challenges management to address new types of uncertainties. These pertain to constraints within the natural environment as well as to institutional action aimed at conserving the natural environment. Based on the literature on perceived uncertainties, we derive six areas of ecology-induced uncertainties. Ecology-driven real options are proposed as a conceptual approach for systematically incorporating these uncertainties into the process of investment appraisals. The paper combines the results in an integrative investment framework and illustrates its application with the case of carbon constraints.

5 Conclusion

Within the debate of the global challenge of sustainable development, natural scientists have investigated the limitations of resource stocks and the carrying capacity of the ecosphere, and protagonists of ecological economics have extended state-of-the-art knowledge such as Boulding's "Spaceship Earth," Daly's "steady state economy," and Meadows's "limits to growth."[16] With respect to carbon, these seminal ideas are gaining practical relevance; carbon is critical for our society and constraints due to its overutilization are emerging. This dissertation considers emerging carbon constraints for strategic management from a firm-level of analysis. In the following concluding remarks, I summarize the general contribution of my work, discuss the results in the broader context of a micro-macro link and management theories, and conclude with an outlook for future research.

Contribution

This dissertation contributes to the organizations and natural environment literature[17] as the phenomena of global warming and energy security are discussed from a strategic management perspective. The arguments are made that firms play a key role in reducing the carbon-dependency of society and, most pointedly, they also will face a more carbon-constrained business environment in the future: On the one hand, price increases of fossil fuels may significantly alter energy market conditions. On the other, tighter climate and energy policies may further change the regulatory environment, and low-carbon efforts of stakeholders may result in an institutional change. However, in order for management to start addressing carbon constraints, new approaches are required, as both carbon dimensions pertain to previously taken-for-granted business conditions, and the timing and the extent of the carbon-related changes are uncertain.

The individual papers contribute on different levels to answering the overall research question of how emerging carbon constraints can be addressed in strategic management. Paper I maps the general challenges ahead and lays the foundation for the entire field of the research. As a key message, the paper concludes that society as a whole as well as firms in particular have to change their current carbon utilization patterns. Papers II and III focus on carbon constraints from an assessment perspective. Paper II defines as a new assessment approach four corporate carbon performance indicators. Among its central messages, it emphasizes that physical and monetary aspects as well as the current and future performance are important when assessing corporate carbon performance. Based on this, paper III derives and applies a framework for corporate carbon performance assessments. The main outcome is a new approach that facilitates carbon-related strategic decisions for firms, policy makers, and on financial markets. Paper IV and V consider managerial implications of carbon constraints. Paper IV discusses the consequences in terms of corporate risks and suggests a new carbon risk management approach for firms. The central message is that carbon increasingly matters for firms on the input and output side; however, different value chains are differently affected. Based on this understanding, paper V derives an investment framework as a new approach for strategic management in order to incorporate ecology-induced uncertainties. The main conclusion is that management has to evaluate different uncertainties regarding the natural environment as well as corresponding institutional changes when determining the profitability of investments.

With respect to the audience for my research, the results have implications for three main users. First, *firms* can utilize the outcomes within their strategic management to scrutinize the relevance of carbon constraints for their business. New insights are generated regarding the emergence of carbon constraints for corporate risk management and how to incorporate them in investment planning. Based on the four indicators, management is able to assess the firm's carbon performance, which is the basis for improving and managing carbon input and output flows. Second, the outcomes facilitate the strategic decision making process of *policy makers*. Significant steps towards a decarbonization of society require industry's significant contribution. In order to evaluate in this context the effectiveness of climate and energy policy measures and to develop new approaches for future policies, the carbon performance indicators can serve as an important information basis. Third, actors on *financial markets* obtain information for their strategic management of loans, assets, and investments. Based on the indicators, investors and financial institutions can assess the extent to which carbon already matters financially and how its relevance for profitability will further increase in the future. Furthermore, they are able to conduct comparative analyses and use the results within their assessment processes, for example for establishing a low-carbon-risk portfolio management.

Discussion

From an industrial ecology standpoint, it is important to acknowledge that carbon constraints influence industrial systems and their carbon flow patterns with respect to a specific scale (it is a global issue affecting almost every value chain) and scope (carbon in- *and* outputs matter), that new management methods are necessary to cope adequately with carbon constraints, and that a sustainable solution for carbon constraints will have to significantly reduce our carbon-dependency. Furthermore, when addressing these aspects, decision makers as well as researchers need to take varying, but often large uncertainties into account: Some carbon constraints may be apparent and already result in economic consequences (e.g., financial losses due to extreme weather events). Others, however, are still less tangible and surrounded by high uncertainty with respect to actual occurrence and severity (e.g. mid-depletion point of oil reserves). In sum, carbon constraints and their uncertainties have implications on three different levels.

On the *macro-level*, there seems to be a growing recognition of the relevance of carbon topics in political agendas: many European governments have enforced renewable energy policies, the EU has started an emission trading scheme, and several states in the United States have adopted similar practices. More decisive action is needed though if the ambitious calls from politicians are to be fulfilled. Notably, this action has to reduce current uncertainties regarding future climate legislation, as this is one of the main arguments being made as to why no sufficient change of carbon-utilization patterns is taking place. From a scientific viewpoint, research has to exhibit national and international carbon flows in order to elucidate and reduce the vulnerability of the economy to changed carbon-utilization patterns. Hence, an uncertainty perspective should be taken when combining models that analyze the carbon flows of nations with economic input–output models. According findings should facilitate policy recommendations to avoid the adverse effects of a sudden emergence of carbon constraints and, furthermore, to pave the way for a more carbon-efficient economy.

On the *meso-level*, carbon constraints influence production systems and, in turn, established production patterns have to be questioned. However, rather than 'just affecting individual firms, this topic bears the potential to fundamentally alter value chains. Not only do the internal activities of firms matter, but the carbon risks taken by suppliers or customers with carbon-intensive operations can also affect the firms' own financial bottom lines. Changes in consumption patterns are an example: if consumers switch their preferences due to emerging carbon constraints, value chains with high carbon intensities will become less competitive than more carbon-efficient ones. Furthermore, interfirm collaboration, as in eco-industrial parks, is a special case of value chain interactions. On the one hand, cooperation based, for example, on combined heat and power generation can foster a more efficient use of carbon resources. On the other hand however, the strong interrelation of firms due to material flow exchanges makes a symbiosis more vulnerable to external shocks. Thus, carbon-related uncertainties should be analyzed in detail when designing interfirm collaborations.

On the *micro-level*, strategic management needs to address a firm's carbon intensity, dependency, exposure, and risk. A life-cycle perspective should therefore be adopted to assess operations with respect to their carbon performance. Among the typical strategies with which to respond to uncertainty spreading risks in a diverse portfolio, maintaining flexibility, and making no-regret moves seem of particular importance. A key challenge is to identify and adopt technological innovations that reduce the carbon dependency of firms. Notably with respect to the related carbon risk (compare paper II and III), an additional challenge due to inherent uncertainties is to identify such technologies that are economically viable in different (uncertain) market environments. This stresses the importance of integrating carbon effects into business case calculations: an early consideration of carbon constraints as a risk factor can also reveal areas for new business opportunities, for example new products or services based on the flexible mechanisms of the Kyoto Protocol.

In light of a management theory standpoint, carbon constraints are important for firms' competitiveness in two dimensions: On the one hand, there is a change of the business environment due to limitations regarding the earth's endowment with natural resources and the carrying capacity of the natural environment. Natural resource endowment is determined by ecological limitations, such as the finite reserves of natural resources within the ecosphere and the depletion of these resources. The carrying capacity addresses the ability of the ecosystem to absorb pollution discharges such as air emissions, and delimits the critical flows of these substances from the anthroposphere to the ecosphere. On the other hand, the business environment also changes as carbon constraints are accompanied by institutional action aimed at conserving the natural environment. This action can be framed as human interventions and responses to carbon issues that constitute both formal and informal pressure on firms. With respect to both dimensions, two main management theory streams seem to be promising for analyzing the competitive effects of carbon constraints on firms and the corresponding corporate responses to address these issues.

From an *organizational* level of analysis, the resource-based view (RBV) allows for a firm-internal consideration of a firm's sustained competitive advantage, i.e. the firm's ability to "conceive of and implement strategies that improve its efficiency and effectiveness".[18] The foundation of this ability builds on a firm's access to physical, human, and organizational resources such as assets, capabilities, organizational processes, or knowledge; if these resources are rare, valuable, and difficult to imitate and to substitute, they constitute a competitively relevant factor.[19] In its original conceptualization the RBV was virtually excluding the natural or biophysical environment. Hart accounts for this by introducing the natural resource-based view.[20] By explaining that a firm's relation to the natural environment can be important for sustained competitive advantage, the natural RBV helps managers and scholars to understand how firms can adequately respond to carbon constraints. This pertains to the input side where firms need to re-conceptualize and re-organize their established carbon-flow and energy-use patterns towards carbon-independent production systems, as well as to the output side where new management approaches are required as the previous competitive environment was not affected by GHG emissions and global warming. In this context, carbon constraints as discussed in this dissertation challenge the implicit RBV assumption that resources are generally available: In fact, fossil fuel scarcity may result in the situation that carbon input resources are absent and climate change is contributing to a continuing change of other resources (such as water or physical locations).[21] To account for a firm's ability to change dynamically and adapt to external changes, Teece, Pisano, & Shuen define dynamic capabilities as "the firm's ability to integrate, build, and reconfigure internal and external competences to address rapidly changing environments".[22] As such, the level of a firm's dynamic capabilities regarding carbon-response strategies determines its success in adapting to a change of the competitive environment induced by carbon constraints. Furthermore, Aragon-Correa and Sharma combine the various ideas described above within a contingency perspective regarding a firm's relation to the natural environment.[23] They argue that certain characteristics of the business environment – uncertainties, complexity, and munificence – moderate the relationship between a firm's dynamic capability of a (proactive) environmental strategy and its competitive advantage. Carbon constraints can affect these characteristics and as such play an important role when analyzing the competitive effects of firms' carbon responsiveness. Paper V discusses the special role of different levels of ecology-induced uncertainties (environmental state, organizational effect, and decision response uncertainty) that are relevant in this context and how these uncertainties affect corporate investment decisions.

Whilst, as illustrated above, the natural RBV allows the investigation of conditions and differences within organizations, there are also developments and effects that result in an overall change of the *organizational fields* an organization is embedded in. At this level of analysis, institutional theory argues that success and competitiveness is affected by the degree to which firms conform to expectations and demands in their institutional environment.[24] Over time, organizations seek to adapt their structures, practices and resource investments in order to gain, maintain, or repair legitimacy, which Suchman defines as "a generalized perception or assumption that the actions of an entity are desirable, proper, or appropriate within some socially constructed system of norms, values, beliefs, and definitions".[25] However, given the fact that all firms within one industry are embedded in the same institutional environment, the internal adaptation processes lead to intra-industry homogeneity, a phenomenon known as isomorphism.[26] Jennings and Zandbergen demonstrate the general usefulness of institutional theory in the context of the natural environment by discussing how consensus is built regarding the concept of sustainability, and how corresponding concepts and practices are developed and diffused among institutions.[27] As such, a systematic analysis of firms' competitiveness with respect to carbon constraints must be embedded in a broader scope of legitimacy, which is driven by "cultured-cognitive, normative, and regulative elements".[28] Scott describes these elements by referring to the three pillars of institutions. First, the cultural-cognitive pillar targets mimetic mechanisms that are based on common beliefs and shared conceptions. In the context of carbon constraints, for instance, it seems to be a given that every (large) firm has to showcase best practice examples with respect to CO_2 or energy efficiency in order to demonstrate that climate change is taken seriously – at least as seriously as other firms. Second, the normative pillar describes normative rules based on values and norms. With respect to carbon constraints an example would be the expectation of Non-Governmental Organizations (NGOs) that firms should reduce their carbon output or should use the Greenhouse Gas Protocol[29] in order to measure their carbon emissions. Third, the regulative pillar considers coercive mechanisms such as governmental regulation, industry standards, or governance structures. Examples in the context of carbon constraints are the European emission trading scheme or carbon input taxes. As such, carbon constraints can be seen as an important source for institutional change.[30] Nevertheless, Hoffman discusses the failure of institutional theory to adequately address the issue of change.[31] Based on his understanding, it is a challenging task to elucidate how organizational fields and institutions within these fields coevolve in the carbon constraints context. As the organizational field defines the available options for a firm's strategic action, new options evolve within an evolving field. Therefore, scrutinizing the future evolution and dominance of Scott's three pillars of institutions will reveal the direction of the carbon constraints induced institutional change and the required adaption measures in light of corporate legitimacy and competitiveness. Eventually, climate change as well as fossil resource availability can be presumed to be relevant for the future institutional change: "an organizational field forms around a central issue".[32]

Future research

As future research, two main literature streams could build on the results of this dissertation and investigate the following subjects in detail. In the *virtue of doing good business* literature,[33] thus far no study explicitly addresses carbon constraints and their linkages to firms' financial performance. In this dissertation, by choosing an instrumental perspective, I go beyond the commonly made argument that firms should react to climate change in terms of their corporate social responsibility. I emphasize that it is the monetary self-interest of firms to actively manage carbon constraints, as this is a new and important determinant of future profitability and competitiveness. However, empirical analyses are needed that prove this assertion and to determine whether this is already the case or if the constraints are still in an emergent phase. Such research could consider different aspects determining the independent variable: With regard to quantitative data, this could be the actual amount of firms' carbon inputs or outputs, the firms carbon intensity, or the carbon flows or carbon intensity in monetary terms, as discussed in paper II. Considering qualitative aspects, corporate carbon management efforts or the sophistication of their climate strategies could serve as a proxy for their carbon performance. Furthermore, my research approach considered carbon constraints mainly in terms of new emerging corporate risks. But whenever risks are adequately addressed, new opportunities emerge for corporate business, too. In connection with this risk-opportunity relationship of carbon constraints, studies could investigate the effects on corporate competitiveness, for example by cross-sectional analyses or different case studies reflecting the business case for a proactive response to carbon constraints.

Considering the literature of *organization and management theory*,[34] scholars could analyze different managerial issues and developments regarding organizations and the natural, carbon-related environment. As described in paper IV, both the carbon input and output flows as well as the direct effects of climate change are important to be addressed by strategic management. But these are two different aspects regarding firms' responsiveness to carbon constraints and their motivation to (re)act. Accordingly, different research questions emerge. On the one hand, mitigation measures pertain to areas such as process optimizations or offsetting activities. These measures might go along with simultaneous cost reductions (e.g., through operational efficiency gains), generate a competitive advantage (e.g., through new customers), or cause additional costs (e.g., by investments in Clean Development Mechanism projects). But most importantly, these measures are responses to a rather continuous change in the business environment.[35] Furthermore, related measures have an immediate effect on a firm or on its performance: Emissions are reduced, products become carbon-neutral, and so on. Therefore, what determines the individual path firms take in terms of their responses to carbon constraints and what are the requirements and conditions for a successful response? What are the necessary firm-specific resources and capabilities necessary to detect and manage the continuous changes in the carbon-related business environment? On the other hand, the need for firms to adapt to global warming increases. Adaptation measures, however, pertain to areas other than mitigation measures. Some firms have to react to a discontinuously changing natural environment,[36] while others might only experience slight, almost irrelevant temperature changes. As a consequence, some firms may be forced to rethink their entire business model, while others may not have to adapt at all. Notably, corresponding measures usually have a rather long-term effect: For example, re-allocated production facilities will most likely reveal their effect only over time, i.e. when it becomes obvious that the previous facility would indeed have been subject to physical damage. Therefore, why do some firms tend to adapt to climate change in a proactive manner while others only respond reactively, and how and why do related firm perceptions differ? What are the resources and capabilities required to assess related risks and opportunities? Finally, both mitigation as well as adaptation measures will be determined by the interactions and pressures of various stakeholders, developments in the natural environment, and institutional changes. Therefore, how do these areas evolve and what do firms require in order to assess the implications for corporate strategy? In sum, this dissertation describes the basic concepts for strategic management under carbon constraints. These can be the starting point for further challenging research questions and elaborate investigations in the field of organizations and the carbon environment.

6 Papers

Each of the papers is included in the annex as originally published by the journal or for those not yet published in the format in which it was submitted to the journal. The submission status of the publications is as of April 30, 2009.

I) **Carbon Constraints in the Fourteenth and Twenty-first Centuries**, with V.H. Hoffmann, 2007. *Journal of Industrial Ecology 11(2)*, 4-6

II) **Carbon Performance Indicators – Carbon Intensity, Exposure, Dependency, and Risk**, with V.H. Hoffmann, 2008. *Journal of Industrial Ecology 12 (4)*, 505-520

III) **Whither Carbonomics? The Carbon Performance of the 100 largest US Electricity Producers**, with G. Weinhofer & V.H. Hoffmann, 2008. *Journal of Industrial Ecology* (submitted)

IV) **Emerging carbon constraints for corporate risk management**, with V.H. Hoffmann, 2007. *Ecological Economics 62 (3-4)*, 518-528

V) **Ecology-driven Real Options: an Investment Framework for Incorporating Uncertainties in the Context of the Natural Environment**, with V.H. Hoffmann, 2009. *Journal of Business Ethics* (in press)

Endnotes

[1] See for example IPCC. 2007. Climate Change 2007: The Physical Science Basis. Cambridge: Cambridge University Press; Lovins, A. B., Datta, E. K., Bustnes, O.-E., Koomey, J. G., & Glasgow, N. J. 2005. Winning the Oil Endgame. Colorado: Rocky Mountain Institute; World Economic Forum. 2008. CEO Climate Policy Recommendations to G8 Leaders. Geneva (http://www.weforum.org).

[2] In the Kyoto Protocol six greenhouse gases are specified, which are measured according to their global warming potential in CO_2-equivalents: Carbon dioxide, methane, nitrous oxide, hydrofluorocarbons, perfluorocarbons, and sulfur hexafluoride. Considering all six under the umbrella of "carbon" is consistent with the other approaches, e.g. the carbon footprint (Carbon Trust. 2007. Carbon footprinting: An introduction for organisations. London: Carbon Trust). Furthermore, CO_2 accounts for the main portion, as on average about 93% of GHG emissions of firms in the FTSE 100 (index of the 100 most highly capitalized firms listed on the London Stock Exchange) are CO_2-related (Henderson, & Trucost. 2005. The Carbon 100 - Quantifying the Carbon Emissions, Intensities and Exposures of the FTSE 100. London: Henderson Global Investors).

[3] Compare Freeman, R. E. 1984. Strategic Management: A Stakeholder Approach. Boston: Pitman; Mitchell, R. K., Agle, B. R., & Wood, D. J. 1997. Toward a theory of stakeholder identification and salience: Defining the principle of who and what really counts. Academy of Management Review, 22(4): 853-886; Sharma, S., & Henriques, I. 2005. Stakeholder influences on sustainability practices in the Canadian forest products industry. Strategic Management Journal, 26(2): 159-180.

[4] Compare Friedman, M. 1962. Capitalism and Freedom. Chicago: University of Chicago Press; Garriga, E., & Mele, D. N. 2004. Corporate Social Responsibility theories: Mapping the territory. Journal of Business Ethics, 53(1-2): 51-71; Jones, T. M. 1995. Instrumental Stakeholder Theory - a Synthesis of Ethics and Economics. Academy of Management Review, 20(2): 404-437.

[5] Compare EIA. 2008. Short-term energy outlook. July 8, 2008 Release, Energy Information Agency (http://www.eia.doe.gov/steo).

[6] See for example Solow, R. 1974. The economics of resources or the resources of economics. American Economic Review 66: 1–114; Hotelling, H. 1931. The economics of exhaustible resources. Journal of Political Economy, 39: 137–175.

[7] See for example the statistics provided by PB (http://www.bp.com/sectiongenericarticle.do?categoryId=9023773&contentId=7044469)

[8] See for example the statistics provided by the International Energy Agency (http://tonto.eia.doe.gov/dnav/ng/ng_pri_sum_dcu_nus_m.htm).

9   Compare for example Basheda, G., Chupka, M. W., Fox-Penner, P., Pfeifenberger, J. P., & Schumacher, A. 2006. Why Are Electricity Prices Increasing? An Industry-Wide Perspective. Washington, D.C.: The Edison Foundation.

10   General information about the emission trading scheme can be found at the European Commission's web page (http://ec.europa.eu/environment/climat/emission/index_en.htm).

11   This can be illustrated, for example, by the automotive firm Volkswagen: in 2005 Volkswagen had to withdraw from the production of the Lupo car, a low-emission car (consuming as little as 3 liters of gasoline per 100 km). The reason for this was a lack of demand (Reed, J. 2007. Problems of pitching cleaner cars to the unconverted. London: Financial Times, Jan 30, 2007). However, in 2008 the firm has pursued a very successful strategy with its Blue Motion line, which consists of a series of low-emission cars.

12   See for example the Carbon Disclosure Project (http://www.cdproject.net).

13   See for example Hirsch, R., Bezdek, R., & Wendling, R. 2005. Peaking of World Oil Production: Impacts, Mitigation, and Risk Management: DOE NETL (http://www.netl.doe.gov/publications/others/pdf/Oil_Peaking_NETL.pdf).

14   EC. 2007. EU action against climate change, Office for Official Publications of the European Communities. Brussels: European Commission.

15   For a general introduction to the concepts of industrial ecology see Graedel, T. E., & Allenby, B. R. 2003. Industrial Ecology (2nd Edition). Upper Saddle River: Prentice-Hall; Ayres, R., & Ayres, L. 2002. Handbook of Industrial Ecology. Cheltenham: Edward Elgar Publishing Ltd.

16   Boulding, K. E. 1966. The Economics of the Coming Spaceship Earth In H. Jarrett (Ed.), Environmental Quality in a Growing Economy: 3-14. Baltimore: Johns Hopkins University Press; Daly, H. E. 1973. Toward a steady-state economy. San Francisco: Freeman; Meadows, D. H., Meadows, D. L., Randers, J., & Behrens, W. W. 1972. The Limits to Growth. New York: Universe Books.

17   Compare the Academy of Management division *Organizations and the Natural Environment* and related publications of its members (http://one.aomonline.org).

18   Barney, J. 1991. Firm Resources and Sustained Competitive Advantage. Journal of Management, 17(1): 99-120, page 102.

19   Barney, J. 1991. Firm Resources and Sustained Competitive Advantage. Journal of Management, 17(1): 99-120; Russo, M. V., & Fouts, P. A. 1997. A resource-based perspective on corporate environmental performance and profitability. Academy of Management Journal, 40(3): 534-559

20   Hart, S. L. 1995. A Natural-Resource-Based View of the Firm. Academy of Management Review, 20(4): 986-1014.

[21] Compare in this context Haigh, N. 2008. Study of Organisational Strategy in Response to Climate Change Issues, PhD Thesis, School of Business, University of Queensland/Australia.

[22] Teece, D. J., Pisano, G., & Shuen, A. 1997. Dynamic capabilities and strategic management. Strategic Management Journal, 18(7): 509-533, page 516.

[23] Aragon-Correa, J. A., & Sharma, S. 2003. A contingent resource-based view of proactive corporate environmental strategy. Academy of Management Review, 28(1): 71-88; compare also Sharma, S., & Vredenburg, H. 1998. Proactive corporate environmental strategy and the development of competitively valuable organizational capabilities. Strategic Management Journal, 19(8): 729-753.

[24] Powell, W. W., & DiMaggio, P. J. 1991. The New Institutionalism in Organizational Analysis. Chicago: The University of Chicago Press; Scott, W. R. 2001. Institutions and Organizations (2nd Edition). London: Sage Publications.

[25] Suchman, M. C. 1995. Managing Legitimacy - Strategic and Institutional Approaches. Academy of Management Review, 20(3): 571-610, page 574.

[26] DiMaggio, P. J., & Powell, W. W. 1983. The Iron Cage Revisited - Institutional Isomorphism and Collective Rationality in Organizational Fields. American Sociological Review, 48(2): 147-160.

[27] Jennings, P. D., & Zandbergen, P. A. 1995. Ecologically Sustainable Organizations - an Institutional Approach. Academy of Management Review, 20(4): 1015-1052.

[28] Scott, W. R. 2001. Institutions and Organizations (2nd Edition). London: Sage Publications, page 48.

[29] WBCSD, & WRI. 2004. The Greenhouse Gas Protocol: A Corporate Accounting and Reporting Standard (Revised version). Geneva, Washington DC: World Business Council for Sustainable Development, World Resources Institute.

[30] Compare to institutional change: Greenwood, R., & Hinings, C. R. 1996. Understanding radical organizational change: Bringing together the old and the new institutionalism. Academy of Management Review, 21(4): 1022-1054; Greenwood, R., Suddaby, R., & Hinings, C. R. 2002. Theorizing change: The role of professional associations in the transformation of institutionalized fields. Academy of Management Journal, 45(1): 58-80; Oliver, C. 1992. The Antecedents of Deinstitutionalization. Organization Studies, 13(4): 563-588.

[31] Hoffman, A. J. 1999. Institutional evolution and change: Environmentalism and the US chemical industry. Academy of Management Journal, 42(4): 351-371.

[32] Hoffman, A. J. 1999. Institutional evolution and change: Environmentalism and the US chemical industry. Academy of Management Journal, 42(4): 351-371, page 351.

[33] Compare for example the discussion in Vogel, D. 2005. The Market for Virtue - The Potential and Limits of Corporate Social Responsibility. Washington, DC: Brookings Institution Press.

34 Compare for general introductions Miles, R. E., & Snow, C. C. 1978. Organizational strategy, structure, and process. New York: McGraw-Hill; Scott, W. R. 2001. Institutions and Organizations (2nd Edition). London: Sage Publications; Thompson, J. D. 1967. Organizations in action. New York: McGraw-Hill.

35 For instance, climate policies and corresponding measures are developed over time (compare the process of establishing the European emission trading system, see footnote 10), consumption patterns do not shift overnight (compare the Volkswagen example in footnote 11), and also financial markets tend to foster their interest in carbon-related issues rather slowly (compare the step-wise evolution of the Carbon Disclosure Project since 2003, see footnote 12).

36 See Winn, M., & Kirchgeorg, M. 2005. The siesta is over: a rude awakening from sustainability myopia. In S. Sharma, & M. Starik (Eds.), Research in Corporate Sustainability, Volume 3, Strategic Capabilities and Competitiveness: 232-258. Northampton: Elgar.

Annex

Paper I

CARBON ACCOUNTING AND DECARBONIZATION

Carbon Constraints in the Fourteenth and Twenty-first Centuries

Volker Hoffmann and Timo Busch

Recently, oil prices have risen to all-time highs, (re)insurance companies are facing more frequent losses due to extreme weather events, and energy-intensive industries are subject to new cost drivers due to carbon dioxide (CO_2) regulation. But are these issues totally new? Yes, as far as the specific details of their occurrence; and no, regarding the underlying problem itself.

A long time ago, Easter Island was covered with lush vegetation. From the fifth century onward, the island's population increased steadily and people created a thriving society. A well-known characteristic of Easter Island is the "moai." These impressive statues were carved during the twelfth to fifteenth centuries, which was the most prosperous time on the island. But when Easter Island was discovered by European sailors in 1722, it was in a morbid condition (Brander and Taylor 1998): there was almost no vegetation, violent conflicts abounded, and cannibalism was part of everyday life. What had happened? Although scientists debate over the details of possible causes, it is common to most explanations that life on Easter Island was very carbon-intense. Wood, with its variety of applications, was critical for the society to function

Carbon is also critical for our society and constraints due to its overutilization are emerging again. But whereas Easter Island "only" suffered local constraints from insufficient wood inputs, today's problem is global in nature and stems from various input and output effects.

(fishing, cooking, housing, and, most notably, transportation of the colossal moai). But by the end of the fourteenth century, most trees on the island had been logged, and over time, this natural carbon resource became depleted. The low availability of carbon constrained the activities of the islanders. As the most visible sign of economic breakdown, the carving of statues stopped first, and later, the entire society collapsed.

Whatever happened on Easter Island, is Diamond (2005) right in stating that our modern society is on a similar track, just with a new scope and on a different scale? Natural scientists have investigated limitations of resource stocks and the carrying capacity of the ecosphere, and protagonists of ecological economics have extended state-of-the-art knowledge such as Boulding's "Spaceship Earth," Daly's "steady state economy," and Meadows's "limits to growth."

With respect to carbon, these seminal ideas are gaining practical relevance: empirically based findings of the Intergovernmental Panel on Climate Change (IPCC) prove that climate change is indeed occurring. Similarly, issues such as the final depletion of crude oil and its mid-depletion point are increasingly being discussed (e.g., Campbell 1997). Hence, carbon is also critical for *our* society and constraints due to its overutilization are emerging again. But whereas Easter Island "only" suffered local constraints from insufficient wood inputs, today's problem is global in nature and stems

© 2007 by the Massachusetts Institute of Technology and Yale University

Volume 11, Number 3

from various input *and* output effects (Busch and Hoffmann, 2007).

From the *input perspective*, carbon constraints are related to the disposition of fossil resources. On the one hand, fossil resources are naturally limited and scarcity is determined by factors such as endowment of natural stocks, access to reserves, and technical developments that postpone final depletion. On the other hand, sociopolitical factors such as taxes, international political developments, and changes in consumer preferences additionally influence the disposition of fossil resources. From the *output perspective*, carbon constraints are related to climate change and can be divided into direct and indirect effects. The former describe direct physical impacts of climate change such as damage to production facilities, availability of raw materials (e.g., water), and impacts on human health. The latter encompass effects of global warming mitigation efforts, such as the European CO_2 Emission Trading Scheme, but also changes in contract conditions (e.g., insurance) and consumer preferences.

From an industrial ecology standpoint, important research questions are whether—or rather, in what way—the scale and scope of today's carbon constraints will influence carbon flow patterns, which management methods are necessary and adequate to cope with carbon constraints, and how meeting our carbon-dependency needs will be affected. Although first tools are already addressing these questions, decision makers as well as researchers need to take large uncertainties into account: Some carbon constraints may be apparent and already result in economic consequences (e.g., losses due to extreme weather events) (Munich Re 2005). Others, however, are still less tangible and surrounded by high uncertainty with respect to actual occurrence and severity (e.g., mid-depletion point of oil reserves). We suggest investigating carbon constraints and their uncertainties on three different levels.

On the *macrolevel*, there seems to be a growing recognition of the relevance of carbon topics in political agendas: many European governments have enforced renewable energy policies, the EU has started emission trading schemes in the EU, and several states in the United States have adopted similar practices. More decisive action is needed, though, if the ambitious calls from President Bush to replace more than 75% of U.S. oil imports from the Middle East by 2025 and from Prime Minister Blair to reduce the U.K. output of CO_2 by 60% by 2050 are to be fulfilled. From a scientific viewpoint, there should be more research to exhibit national and international carbon flows in order to elucidate and reduce the vulnerability of the economy to changed carbon utilization patterns. Hence, an uncertainty perspective should be taken when combining models that analyze the carbon flows of nations (e.g., Uihlein et al. 2006) with economic input–output models. The findings can be a source for policy recommendations to avoid adverse effects of a sudden emergence of carbon constraints and, furthermore, to pave the way to a more carbon-efficient economy.

On the *mesolevel*, carbon constraints influence the business environment and, in turn, established production patterns will be questioned. But rather than "just" affecting individual companies, the topic bears the potential to fundamentally alter value chains: Not only do the internal activities of companies have to be assessed, but also carbon risks of suppliers or customers with carbon-intensive operations can affect the companies' own financial bottom lines. Changes in consumption patterns are an example: if consumers switch their preferences due to emerging carbon constraints, value chains with high carbon intensity will become less competitive than more carbon-efficient ones. Furthermore, interfirm collaboration, as in eco-industrial parks, is a special case of value chain interactions. On the one hand, cooperation based, for example, on combined heat and power generation can foster a more efficient use of carbon resources. On the other hand, however, the strong interrelation of firms due to material flow exchanges makes a symbiosis more vulnerable to external shocks. Hence, carbon-related uncertainties should be analyzed in detail when designing interfirm collaborations.

On the *microlevel*, strategic management needs to assess and address a company's carbon exposure. A life-cycle perspective should be adopted to evaluate operations with respect to their intensity of and dependency on carbon. Among the typical strategies to respond to uncertainty—for example, avoid, reduce, adapt—spreading risks in a diverse portfolio, maintaining flexibility, and making no-regret moves seem of particular importance. A key challenge is to identify and adopt technological innovations that reduce the carbon exposure of firms. An additional challenge due to inherent uncertainties is to identify such technologies that are economically viable in different (uncertain) market environments. This stresses the importance of integrating carbon effects into business case calculations: an early consideration of carbon constraints as a potential risk factor can also reveal areas for new business opportunities, for example, new products or services based on the flexible mechanisms of the Kyoto Protocol.

We believe that even though Easter Island was subject to its very own carbon constraints, early action in terms of adaptation (e.g., adjustments to weather-related risks) and mitigation (e.g., accelerated exploitation of efficiency gains) can help turn emerging carbon constraints of the twenty-first century from an economic risk into a business opportunity. And, most important, the necessary insights and measures can help prevent carbon constraints from again becoming a severe issue, notably on a global scale.

References

Brander, J. and S. Taylor. 1998. The simple economics of Easter Island: A Ricardo–Malthus model of renewable resource use. *American Economic Review* 88(1): 119–138.

Busch, T. and V. H. Hoffmann. 2007. Emerging carbon constraints for corporate risk management. *Ecological Economics* 62(3–4): 518–528.

Campbell, C. 1997. *The Coming Oil Crisis*. Essex, UK: Multi-Science Publishing.

Diamond, J. 2005. *Collapse: How Societies Choose to Fail or Succeed*. Viking.

Munich Re. 2005. *Topics Geo, Annual Review: Natural Catastrophes 2004*. Munich: Munich Reinsurance Group.

Uihlein, A., W. F. Poganietz, and L. Schebek. 2006. Carbon flows and carbon use in the German anthroposphere: An inventory. *Resources Conservation and Recycling* 46(4): 410–429.

About the Authors

Volker Hoffmann is an assistant professor and **Timo Busch** is a Ph.D student at the Swiss Federal Institute of Technology (ETH) in Zurich.

Address correspondence to:
Professor Dr. Volker Hoffmann
Group for Sustainability and Technology
ETH Zurich
8092 Zurich, Switzerland
<vhoffmann@ethz.ch>
<www.sustec.ethz.ch>

Annex

Paper II

FORUM

Corporate Carbon Performance Indicators

Carbon Intensity, Dependency, Exposure, and Risk

Volker H. Hoffmann and Timo Busch

Keywords:

carbon management
climate change
environmental indicator
fossil fuel
greenhouse gases (GHGs)
industrial ecology

Summary

The dependency on carbon-based materials and energy sources and the emission of greenhouse gases have been recognized as major problems of the 21st century. Companies are central to the effort to grapple with these issues due to the large material flows they process and their capabilities for technological innovation. It is important, on the one hand, to determine the individual stake companies have in these issues and, on the other, to measure companies' performance. Since the results of studies thus far have been ambiguous, we define four comprehensive and systematic corporate carbon performance indicators: (1) Carbon intensity is physically oriented and represents a company's carbon use in relation to a business metric. (2) Carbon dependency illustrates the change in physical carbon performance within a given time period. (3) Carbon exposure reveals the financial implications of using and emitting carbon. (4) Carbon risk estimates the change in financial implications of carbon usage within a given time period. On the basis of these general definitions, we specify the indicators for a standardized application that can support two important stakeholders in their decision making: policy makers, who can include such information when evaluating current climate policies and formulating future ones, and investors and financial institutions, which can compare companies with respect to their carbon performance and corresponding financial effects.

Address correspondence to:
Prof. Dr. Volker H. Hoffmann
ETH Zurich
Department for Management, Technology and Economics
Kreuzplatz 5
8032 Zurich, Switzerland
vhoffmann@ethz.ch
www.sustec.ethz.ch

© 2008 by Yale University
DOI: 10.1111/j.1530-9290.2008.00066.x

Volume 12, Number 4

FORUM

Introduction

The world faces twin energy-related threats: "that of not having adequate and secure supplies of energy at affordable prices and that of environmental harm caused by consuming too much of it" (IEA 2006, 1). With respect to the supply side, the availability of all fossil resources is naturally limited in the long run, and crude oil, as one key carbon input for economies, is thought by many to be about to peak (Bentley 2002; Deffeyes 2003; Campbell 2005). As a result, further price increases for carbon-based inputs are inescapable in the long run (Reynolds 1999; IEA 2006). With respect to the consumption side, global warming will have drastic ecological consequences (IPCC 2007a) and possibly far-reaching economic implications (Stern 2006). Policy makers have started to take up these challenges: For example, the European (EU) has set a greenhouse gas (GHG) reduction target of 20% for 2020 and developed the vision to decarbonize society by 60% to 80% by 2050 (EC 2007b).

Companies are central to paving the way toward a low-carbon society, because a large portion of carbon inputs and GHG emissions stems from industrial production. As a consequence, stakeholders increasingly require companies to disclose their strategies for addressing climate change. In particular, actors in financial markets are investigating the implications of climate change and corporate responses on the competitive position of companies and on risks to shareholder value (e.g., CERES 2006; Innovest 2006). However, business responses to climate and carbon issues have been characterized as ambiguous, and external assessments of corporate efforts have been contradictory, even when the same firms are analyzed (Jones and Levy 2007). Furthermore, for many companies, emissions from their own operations are dwarfed by emissions that occur upstream or downstream in the value chain—for example, those connected to energy provision or product usage, which are often not covered in voluntary GHG emission reports. To increase the reliability of life cycle–wide carbon assessments and to determine performance differences between companies, researchers must have indicators that concisely measure a company's performance with respect to carbon.

Industrial ecology literature stresses that accounting for carbon-based materials and GHGs has always been an important part of life cycle analysis and modeling (Lifset 2007). For example, Morioka and colleagues (2005) use carbon dioxide (CO_2) emissions as an indicator to design advanced loop-closing systems for the recycling of end-of-life vehicles and electric household appliances. With respect to products and services, the carbon footprint method has become a dominant method (EC 2007a): The Climate Footprint Calculator[1] and analyses of carbon footprints in supply chains (Carbon Trust 2006a) are recent examples. Regarding organizations, the Global Reporting Initiative (GRI 2006) defined broadly applied reporting standards that also include carbon emissions, whereas the World Business Council for Sustainable Development and the World Resources Institute (WBCSD and WRI 2004) developed a sophisticated accounting method for GHGs. Nonetheless, neither the scientific nor the practitioner-oriented literature contains a consistent set of indicators that also includes carbon input materials and that relates the way companies use carbon to their underlying business activities. Rather, different definitions and interpretations of the same expressions abound, and there is no common understanding of how to report or analyze a company's use of carbon or emission of GHGs.

To help in filling this gap, in this article we aim to increase the transparency of corporate carbon performance assessments by defining four comprehensive and systematic indicators for analysis and reporting purposes. The indicators shed light on the physical and monetary dimensions of a company's current and future activities with respect to carbon inputs and outputs. We suggest a specification for practical application of these indicators that enables stakeholders to assess a company's stake in climate change and its efforts toward better managing carbon usage: Policy makers can use such information to formulate and evaluate policies, whereas financial markets can obtain insights regarding the performance of companies with respect to carbon and corresponding financial effects.

FORUM

A Company's Link to Carbon

We refer to the extent to which a company's operations and its value chain are based on carbon as *carbon usage*. A company's carbon usage comprises (1) an input dimension that relates to production processes that utilize carbon-based materials and energy and (2) an output dimension that refers to the emission of GHGs from these production processes (Busch and Hoffmann 2007). Although not all GHGs directly relate to carbon, we include them in our considerations in terms of their CO_2 equivalents.[2]

System Boundaries and Scope of Carbon Usage

The carbon usage of a company depends on the industry it operates in, its position in the value chain, and company-specific factors, such as product portfolio or technological equipment. For a comprehensive analysis of a company's carbon usage, a cradle-to-grave perspective is important (cf. Burner et al. 2008), and direct and indirect carbon inputs and outputs have to be distinguished (see figure 1). For the output dimension, the Greenhouse Gas Protocol Initiative (WBCSD and WRI 2004) developed a classification scheme. This approach is sufficient for analysis of the output dimension of corporate carbon usage. For analysis of monetary implications, however, carbon inputs also matter. Therefore, we extend this scheme and describe different system boundaries as three scopes for direct and indirect levels of carbon usage in the input and output dimensions (see table 1).

Within the gate-to-gate view of scope 1, only direct carbon usage is taken into account, and neither upstream nor downstream aspects are considered. The determination of scope 1 carbon usage requires the least effort in terms of data gathering and analysis. The results display only a very limited part of the actual carbon usage, however, excluding, for example, "gray" carbon usage that occurs during the production of supplied products. In contrast, the combined consideration of scopes 1 and 2 also includes the carbon usage relating to purchased energy. Nonetheless, negative financial effects can still lurk behind other parts of the value chain—for example, when suppliers pass on emission costs to their customers. To address this problem, one can analyze the full value chain, including the upstream carbon usage of the company's supply chain as well as the downstream carbon usage linked to a company's products and services (scope 3). Nevertheless, the determination of this scope is shaped by practical limitations, such as data availability and the time required for precise and reliable data analyses.

Measuring Corporate Carbon Performance

Thus far we have centered our remarks on the absolute carbon usage of companies (e.g., the total amount of GHG emissions). This is important for general or aggregated trend

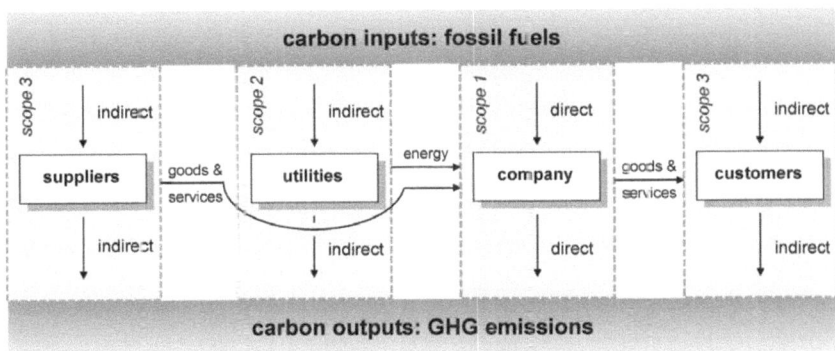

Figure 1 Carbon inputs and outputs with varying system boundaries.

FORUM

Table 1 Carbon usage within three scopes

Scope	Carbon input dimension	Carbon output dimension (GHG protocol)
Scope 1	Direct carbon input: • Used as material component within on-site production processes • Used for direct combustion of fossil fuels in boilers and furnaces • Used as energy source for on-site power generation	Direct GHG emissions: • From on-site production processes • From direct combustion of fossil fuels in boilers and furnaces • From on-site power generation
Scope 2	Indirect carbon input: • Used as energy source for purchased energy (electricity, heat, steam)	Indirect GHG emissions: • From consumption of purchased energy (electricity, heat, steam)
Scope 3	Other indirect carbon inputs (not included in Scope 1 or Scope 2): • Required for or within upstream and downstream processes • Associated with outsourced or contracted activities • Other inputs	Other indirect GHG emissions (not included in Scope 1 or Scope 2): • From upstream and downstream sources • Associated with outsourced or contracted activities • Other emissions

Source: Based on WBCSD and WRI (2004).
Note: GHG = greenhouse gas.

investigations with a sectorwide or macroeconomic view that can help to identify situations in which individual companies or industry sectors continuously increase their absolute emissions even while governments pursue national emission reduction goals. Nonetheless, to compare the carbon usage across companies and to incorporate changes in a company's business activities over time (e.g., through mergers and acquisitions), it is important to place the absolute carbon usage in relation to a business metric.

The generation of this type of ratio has been thoroughly discussed in the ecoefficiency literature (e.g., Schaltegger and Sturm 1990; WBCSD 2000). Ecoefficiency measures illustrate the economic output that is obtained from a given resource input or that generates a given environmental effect (DeSimone and Popoff 1997;

Table 2 Business metrics relevant to carbon indicators

Business metric	Description
Unit of production	Business output in physical units; no consideration in monetary terms
Turnover (or sales)	Value of the company's production step in the value chain plus all upstream business activities; considers cradle-to-gate value creation
Total costs	Expenses for generating the business output; considers company's costs, including all expenses in the profit and loss statement
Costs of goods sold	Expenses that exclude indirect costs, such as office costs; shows direct expenses incurred in producing the company's output
Value added	Sales less intermediate costs for purchased goods and services; emphasis is put on the company's production step within the value chain
Earnings before interest and taxes (EBIT)	Approximate measure of a company's operating cash flow; focal point is the profitability of the company
Market capitalization or equity	Market value of a company or value of equity; emphasis is put on the value of the company as a whole

Source: Our own compilation, based on Horngren and colleagues (2006) and Henderson and Trucost (2005).

	static approach	*dynamic approach*
physical units	carbon intensity	carbon dependency
monetary units	carbon exposure	carbon risk

Figure 2 Corporate carbon performance indicators.

Burritt and Saka 2006).[3] In contrast, the inverted ratio, ecointensity (Bartelmus et al. 2004; Ehrenfeld 2005), considers the amount of environmental impact in relation to a business metric. Depending on the business metric chosen (see table 2), the ecointensity metric can have different explanatory powers. The advantage of using an intensity measure instead of an efficiency measure to illustrate corporate carbon performance is that comparisons between companies and reduction potentials become more transparent.

Considering a company's carbon usage and the delineated business metrics, we propose four corporate carbon performance indicators (see figure 2): First, we take a static view and analyze the physical carbon performance from a material flow perspective (carbon intensity). Second, we again analyze physical flows but take a dynamic view by considering how much the company relies on carbon over time (carbon dependency). Third, we go beyond the purely physical carbon flows and analyze the monetary implication of carbon intensity from a static perspective (carbon exposure). Fourth, we combine the dynamic view regarding carbon dependency with monetary implications and discuss how to derive a corporate risk figure that allows conclusions on carbon's financial importance over time (carbon risk). We now describe each indicator in detail.

Determining Carbon Intensity

The term *carbon intensity* has been used in various ways in the literature as well as in practical applications. On the macro level, approaches have considered carbon inputs (EIA 1995; Bosetti et al. 2006; Huesemann 2006) as well as carbon outputs (Lebel et al. 2007; Raupach et al. 2007) to determine intensity indicators that describe carbon flows within an economic system. On the micro level, carbon intensities are used for the internal or external analysis of companies, for reporting purposes, and for ranking different companies (for examples, see table 3). The current usage of these indicators is problematic for four reasons, however: Different synonyms abound for the same underlying indicator, the same synonyms are used for different underlying indicators, system boundaries vary among scopes 1–3, and carbon intensities are only based on carbon outputs, not on carbon inputs. This makes it very difficult for external stakeholders to compare the carbon intensities of different companies.

To derive a consistent terminology, we suggest a general definition of the term *carbon intensity*.

Definition 1: *Carbon intensity relates to a company's physical carbon performance and describes the extent to which its business activities are based on carbon usage for a defined scope and fiscal year.*

The intensity is measured by the ratio of a company's carbon usage in absolute terms to a related business metric. The carbon usage specifies the amount of carbon the company utilizes or emits for a chosen scope and fiscal year. The business metric is a measurement of a company's financial performance for the same fiscal year. In the determination of a company's carbon intensity, the time frame and the material flow level are relevant.

Table 3 Differences in use of the term *carbon intensity*

Objective	Indicator	Carbon usage	Business metric	Source
Tracking and reporting on emission reductions	Metrics for measuring GHG emissions	CO_2-eq (in tons; scope varies)	Unit of product (no specification)	Hoffman (2006)
Comparison of different kinds of fossil fuels	Emission coefficient	CO_2 (in pounds; scope 1)	Fuel unit (volume or mass) or British Thermal Units	EIA (2007)
Ranking of investment funds	Carbon intensity	CO_2-eq (in tons; scope 1–3)	Market capitalization (in £ or other currencies)	Trucost (2006)
External analysis of companies	GHG emission intensity	CO_2-eq (in tons, scope 1–3)	Unit of production (no specification)	Maxime and colleagues (2006)
	Carbon intensity	CO_2 (in kg; scope 1)	Unit of production (in MWh)	Voisin and Lamotte (2006)
	Carbon intensity	CO_2 (in g, scope 1)	Unit of production (in kWh)	Hutchinson (2006)
	Carbon intensity	CO_2-eq (in tons; scope 1–3)	Turnover, EBITDA, or market capitalization (in £)	Henderson and Trucost (2005); Societe Generale (2007)
Internal analysis and reporting	GHG intensity	CO_2-eq (in tons; scope 1)	Unit of production (thousand barrels of oil equivalent)	BP (2006)
	GHG emissions intensity	CO_2-eq (in tons; scope 1–2)	Unit of production (cubic meters of oil equivalent)	Encana (2007)
	Production carbon intensity	CO_2-eq (in tons; scope 1–2)	Unit of production (cubic meters of oil equivalent)	Petro-Canada (2007)
	CO_2-eq emission intensity	CO_2-eq (in kg; scope 1, partly scope 2–3)	Sales (in US$)	Brambles (2006)
	Key figure for GHG emissions	CO_2-eq (in tons; scope 1–3)	Sales (in CHF)	Roche (2007)
	Emissions volume per sales unit	CO_2-eq (in tons; scope 1)	Sales (in 100 Yen)	Toyota (2006)
	Specific emission data	CO_2 (in kg; scope 1)	Unit of production (cementitious materials in tons)	Holcim (2005)
	Environmental performance indicator	CO_2 (in kg; scope 1)	Unit of production (in tons)	Nestlé (2007)
	Carbon intensity	CO_2 (in tons; scope 1)	Unit of production (in MWh)	E.on UK (2006)

Note: GHG = greenhouse gas; CO_2-eq = carbon dioxide equivalent; CO_2 = carbon dioxide; EBITDA = earnings before interest, taxes, depreciation, and amortization; CHF = Swiss Francs.

With respect to the time frame, the status quo (t_o) and the predicted (t_1) carbon intensity can be distinguished. The *status quo carbon intensity* is based on verified data (e.g., of the previous fiscal year) and as such provides a realistic picture of the company's current carbon intensity. It conforms to the established use of this indicator, as exhibited in table 3. The *predicted carbon intensity* relates to a future time t_1 and is commonly not reported. It requires specification of three parameters: First, a time period has to be defined over which the analysis of the company's carbon performance will be conducted. Second, future price and market conditions have to be estimated on the basis of forecasts such as those issued by the International Energy Agency (IEA). These should specify carbon-related conditions of the business environment and incorporate all relevant information known or assumed to influence the carbon and energy market (notably, future carbon prices; cf. table 4). Third, the company-specific carbon usage in the future has to be estimated. Reflecting the assumptions regarding the future price and market conditions, technological options and alternative production processes have to be identified that a company is likely to use to realize carbon optimization potentials. These potentials relate to efficiency increases (e.g., energy efficiency) and substitution options (e.g., using renewable energy sources instead of fossil fuels). The extent to which a company is able to take advantage of these potentials is determined by the individual status of a company's technology equipment, current government policies, and other factors.

Given the material flow level, a company's carbon input or carbon output intensity can be distinguished, depending on the way carbon usage is calculated. The carbon input intensity relates to the amount of carbon that is needed within the production process. For example, in the plastics industry, the carbon flows can be included that are required for the production of polymers and do not produce emissions. In contrast, the carbon output intensity accounts for the emission of GHGs and acknowledges the company's internal and external efforts to curb emissions via measures such as carbon offsetting (WBCSD and WRI 2004).

The amount of carbon inputs C_I in tons of carbon is calculated as $C_I = \sum_{k=1}^{K_I} C_{I_{k,t}}$ where $k = 1, \ldots, K_I$ is the index for the K_I different inputs and t is the fiscal year of analysis. The scope 1 input data can be obtained from an input-output analysis or derived from the company's accountancy or controlling systems. The determination of scope 2 carbon inputs requires the amount of purchased energy and information regarding the specific energy mix—that is, the carbon inputs for producing the purchased energy—which is available in standard databases, such as the ecoinvent database.[4] For a complete allocation of scope 3 carbon inputs, a life cycle assessment is usually necessary (e.g., Ardenti and Gilardi 2007; Weidema et al. 2008). The carbon input intensity ($C_I In$) can be derived for a chosen scope $i = 1, 2, 3$ and fiscal year t when a business metric (BM) is taken into account.

$$C_I In_{i,t} = \frac{\sum_{k=1}^{K_I} C_{I_{k,t}}}{BM} \quad (1a)$$

The carbon output intensity is based on a company's GHG emissions, measured in CO_2 equivalents and denoted by $k = 1, \ldots, K_O$. Scope 1–3 emissions can be obtained from the same sources as described for carbon inputs. In addition, carbon-relevant activities, such as GHG reductions via offsetting, have to be taken into account. The carbon output intensity ($C_O In$) can be derived analogously to equation (1a).

$$C_O In_{i,t} = \frac{\sum_{k=1}^{K_O} C_{O_{k,t}}}{BM} \quad (1b)$$

Determining Carbon Dependency

In the literature, the notion of carbon dependency has been treated on a macroeconomic level (e.g., Liberatore 2001). It is related to the energy dependency of nations, which refers to the dependency on external energy resources, such as the EU's dependency on Russian natural gas (Kuik 2003). Carbon dependency can be interpreted in the sense of such an energy dependency but also as one country's dependency on another for meeting emission reduction targets through the importation of low-carbon fuels or emissions trading (Wieczorek 2003). In this way, carbon dependency is a "carbon (trading) dependency" (Kuik 2003, 236). We transfer this idea to the micro level and extend it by introducing a time component.

Definition 2: *Carbon dependency describes the change in a company's physical carbon performance within a given time period. The indicator is measured as the company's relative performance change from the status quo to the predicted carbon intensity.*

A company's carbon dependency indicates what percentage of the current status quo (t_0) carbon intensity will remain, with the presumption that the company pursues its business under the assumptions made for estimating its predicted carbon intensity (t_1) in the three steps (above). If a company undertakes all economically feasible efforts to reduce its carbon intensity under these assumptions, the carbon dependency describes the degree to which the company is able to reduce its carbon intensity. As a result, a highly carbon dependent company is hardly able to reduce its carbon intensity over the considered time period. Given the same scope $i = 1, 2, 3$ for both carbon intensities (t_0 and t_1), the carbon dependency (CDe) is expressed as a percentage of the t_0 carbon intensity for the time period $t = t_1 - t_0$.

$$C_I De_{i,\ t} = \frac{C_I In_{i,t_1}}{C_I In_{i,t_0}} \times 100 \quad \text{or}$$

$$C_O De_{i,\ t} = \frac{C_O In_{i,t_1}}{C_O In_{i,t_0}} \times 100 \quad (2)$$

Determining Carbon Exposure

The term *carbon exposure* is often used in practitioner-oriented reports. Carbon Trust (2006b), Henderson and Trucost (2005), and Societe Generale (2007) deliver ratio-based definitions that relate current carbon outputs to different hypothetical future cost figures. As a result, their metrics mix two time frames when carbon usage of t_0 is related to carbon prices of t_1. Furthermore, these authors utilize different business metrics and scopes for carbon usage. Taking a broader view, Schultz and Williamson (2005) also recommend including diverse cost effects relating to customer and shareholder sentiments and climate change events (e.g., rising sea levels) in assessments of a company's carbon exposure. In sum, there is neither a clear distinction between different time periods nor a common understanding of the underlying parameters for assessing companies' carbon exposure. Therefore, we suggest a combined consideration of the input and the output dimensions of a company's carbon usage in one monetary term for one point in time.

Definition 3: *Carbon exposure relates to a company's monetary carbon performance and describes the monetary implications of the business activities due to carbon usage for a defined scope and fiscal year.*

The exposure is measured by the ratio between a company's carbon usage in monetary terms and a related business metric. Through the use of prices, the two ratios that were necessary on the material flow level (i.e., carbon input and output intensity) can be combined in one monetary figure (carbon exposure). One calculates the carbon usage in monetary terms by applying the input prices p_{Ik} with $k = 1, \ldots, K_I$ for each unit C_{Ik} and the output prices p_{Ok} with $k = 1, \ldots, K_O$ for each unit C_{Ok}. Carbon input prices are determined by the mass fraction of each carbon-containing input (e.g., the extent to which products are composed of specific carbon inputs) and the expenditures associated with the initial carbon source (e.g., costs for crude oil and related carbon taxes). On the basis of equation (1), a company's carbon exposure (CEx) can be derived for a fiscal year t.

$$CEx_{i,t} = \frac{\sum_{k=1}^{K_I} C_{I_k,t} \times p_{I_k,t} + \sum_{k=1}^{K_O} C_{O_k,t} \times p_{O_k,t}}{BM}$$
(3)

Similar to carbon intensity, the status quo (t_0) and the predicted (t_1) carbon exposure can be distinguished. For a company's *status quo (t_0) carbon exposure*, we differentiate between two cost approaches, one representing a company-internal perspective and the other a market perspective. The company-internal perspective considers actual costs of fossil fuels and purchased energy during one fiscal year (t_0) as obtained from the company's cost accounting system. To determine the cost of the fossil fuel–related purchased energy, one has to incorporate the country-specific energy mix into the calculations (the percentage of

Table 4 Cost approaches for determining future carbon prices

Approach	Costs to be determined for	Source (e.g.)
Price scenarios	Carbon inputs	Scenarios or forecasts: IEA (2006) or EIA (2008)
	Carbon outputs (GHG)	Scenarios or forecasts: Trexler (2005), Urdal and colleagues (2006), Carbon Trust (2006b), Trucost (2006), or Hourcade and colleagues (2007)
Abatement costs based on mitigation options	Efficiency increase and fossil fuel substitution	Industry specifications: Metz and colleagues (2001), Llewellyn (2007), or Vattenfall (2007)
		Estimates for fossil fuel alternative energy technologies: Barker and colleagues (2006)
	Offsetting	Average prices per ton via CDM project: Capoor and Ambrosi (2007)
		Estimates for CCS: IPCC (2005) or Anderson (2006)
Cost of internalizing external effects	Depletion of fossil fuels	External costs of oil consumption: Sabour (2005)
		Losses due to the depletion of resources such as oil: Weitzman (1999)
	Damage related to carbon emissions	Social cost of carbon emissions: Clarkson and Deyes (2002)
		Optimal carbon price: Nordhaus (2006)

Note: GHG = greenhouse gas; CDM = clean development mechanism.

carbon-based energy production). Similarly, carbon output costs due to, for example, the European Emission Trading Scheme (EU ETS) or voluntary emission reductions (e.g., offsetting via clean development mechanism [CDM] projects) have to be accounted for. In contrast to this company-internal perspective, the use of carbon cost based on market prices follows an opportunity cost logic, and company-specific price conditions are not taken into account. The advantage of utilizing carbon costs based on market prices is that for comparative analyses of different companies, only one price for each input and output has to be determined, which can be obtained from public sources.[5] This reduces complexity and facilitates the monetary assessments involved when more than one company is analyzed.

To estimate the *company's predicted* (t_1) *carbon exposure*, one has to take into account the same carbon-related conditions of the business environment as applied for the t_1 carbon intensity. For the future carbon prices ($p_{Ik,t1}$ and $p_{Ok,t1}$) for the carbon inputs ($C_{I,t1}$) and outputs ($C_{O,t1}$), we distinguish among three approaches (see table 4).

First, scenarios for future prices can be applied, which have to take into account regulatory and market risk and can be based on one's own analyses or on forecasts in the literature. Notably, the scenarios must account for typical correlations between market prices of fossil fuels and CO_2 allowances (Bailey 1998; Montero and Ellerman 1998; Voisin and Lamotte 2006). An estimation of future costs of carbon (CO_2 equivalents) has to be included, for example, when an emission-trading scheme is likely to be in place. One can do this either by choosing an opportunity cost perspective or by assuming a fixed percentage of grandfathered allowances. In the former case, the same price is applied for all GHG emissions; in the latter case, a price would be considered only for a certain percentage of all emissions. Allowance prices have fluctuated considerably in the past and are likely to remain volatile in the future. Forecasts for future GHG price levels are generally difficult, as prices will depend on not yet fully specified variables, such as evolving climate policy (Trexler 2005). One the one hand, forecasts can be derived from historic market prices or forwards for EU ETS allowances, but they should further include a scenario component. One can obtain this component by following a dynamic approach—for instance taking into account decision points in the international policy process

FORUM

and historic price volatilities. Alternatively, to reduce complexity, one could choose a linear approximation assuming a steady price increase (e.g., due to inflation). On the other hand, GHG price forecasts can take into account policy scenarios, resulting effects on carbon markets, and further variables, such as market psychology. Notably, this approach is relevant for companies when they are incorporating future GHG prices in strategic management by determining a "best available corporate forecast" (Trexler 2005, 12).

Second, sector-specific abatement costs can be applied that are associated with corporate measures to curb carbon usage. Basically, three options exist to reduce a company's carbon usage: efficiency increases in existing production processes, substitution away from fossil fuels to become independent of carbon resources, and offsetting strategies to compensate emissions (Weinhofer and Hoffmann, forthcoming). When one is applying abatement costs to determine future carbon costs, it is important to acknowledge that corresponding measures not only generate costs (e.g., by requiring investments) but also reduce costs (e.g., by increasing efficiency) or even might generate additional revenues (e.g., by an excess of allowances). Therefore, the dynamics between additional incurred costs and resulting cost savings and revenues have to be taken into account.

Third, external costs can be included in a long-term business perspective (Scholz and Wiek 2005). Costs of external effects are usually calculated on the macroeconomic level (e.g., Peskin and Angeles 2001); their application on the company level follows the "polluter pays" principle. With respect to carbon inputs, there are efforts in the economic literature to attach a price to the depletion of fossil fuels (e.g., Weitzman 1999). With respect to carbon outputs, cost due to external effects of emitting GHGs can be determined by the cost–benefit or the marginal cost approach (Clarkson and Deyes 2002). Both approaches utilize a figure for the damage per ton of carbon emitted.

Determining Carbon Risk

In general, the term *carbon risk* is often used to describe any corporate risk related to climate change or the use of fossil fuels. Most of the practitioner-oriented reports cited thus far do not explicitly distinguish among carbon exposure, carbon risk exposure, and carbon risks. As a clear definition, Urdal and colleagues (2006) use the term *value at risk from carbon* and calculate the effects of different price scenarios for CO_2 on the equity value of an energy utility. Also, Carbon Trust (2006b) measures the risk from climate change in terms of a carbon exposure (a price for emissions is applied) and further regulatory and market dynamics as well as broader climate change impacts. The resulting risk value is expressed as a percentage of the earnings before interest and taxes (EBIT). We build on this understanding of carbon risk as a foresight indicator but emphasize that carbon risk describes the likely change in carbon-related monetary implications for a company, which one can obtain by determining a company's carbon exposure for both time periods separately and comparing the change between t_1 and t_0.

Definition 4: *Carbon risk describes the change in a company's monetary carbon performance within a given time period. The indicator is measured as the relative performance change from the status quo to the predicted carbon exposure.*

A company's carbon risk indicates by the what percentage current status quo (t_0) carbon exposure will change if the company pursues its business under the assumptions for the predicted carbon intensity (t_1). On the basis of the carbon price scenarios designed for determining the predicted carbon exposure (t_1), the carbon risk displays how the relative monetary relevance of carbon is likely to decrease or increase for the company. As such, it facilitates the comparison of the monetary implications of different companies' carbon usage over time. As a result, a company with a high carbon risk will face a significant increase in the relevance of its carbon performance for the company's costs and profits over the considered time period. If we assume the same scope $i = 1, 2, 3$ for both carbon exposures (t_0 and t_1), the resulting carbon risk (CRi) is derived for the time period $t = t_1 - t_0$.

$$CRi_{i,\ t} = \left(\frac{CEx_{i,t_1}}{CEx_{i,t_0}} - 1 \right) \times 100 \qquad (4)$$

Specification of the Performance Indicators

The four indicators describe different areas that are important in the analysis of a company's carbon performance. The choice of scope of analysis, business metrics, and cost approach determines the explanatory power of the results. Due to the multitude of options pursued in the literature, there is currently a lack of transparency regarding the applied methods. Comparative analyses are impeded, as companies as well as analysts use different approaches arbitrarily. Therefore, we specify our general definitions and suggest one specific set of indicators that appears most promising as a standardized approach to the practical application of the external analysis of companies (table 5). Data availability can be limited, however—for example, regarding scope 3 carbon usage or company-specific costs. Therefore, our approach represents a trade-off between the public availability of data and the requirements for a standardized application of comprehensive indicators.

To determine the carbon intensity, we suggest the consideration of carbon outputs, as this indicator points to corporate efforts in terms of offsetting as well as internal measures, such as efficiency increases. Furthermore, we suggest a focus on scope 1–2, as the determination of scope 3 remains complex (Carbon Trust 2007). The carbon exposure indicator in t_0 relies on market prices, because this significantly reduces data-collecting efforts in analysis of different companies. For deriving the carbon output cost for t_0 carbon exposure we suggest to consider the actual costs for emission reduction certificates related to scope 1 GHG emissions. For all other emissions we suggest taking an opportunity cost perspective. This entails considering a price in t_0 for all scope 2 GHG emissions if an emission trading is in place and a price in t_1 for all scope 1–2 GHG emissions; potential effects of grandfathering and other effects, such as the actual amount of required allowances in t_1 are neglected. Furthermore, we suggest taking the average price of EU ETS allowances of t_0 and applying an annual inflation rate to approximate for the future carbon price.

An Example

To illustrate the practical application of the suggested indicators, we analyze two hypothetical companies that face identical starting points. Both are industrial companies based in the United States that generate their own GHG emissions (scope 1) and require externally produced energy (scope 2). As both companies produce a variety of products, we chose sales as business metrics and assume an annual growth in production and sales of 5%. We assess

Table 5 Specification of a standardized approach for practical application of the four indicators

Required data	Suggested approach
Business metrics	*Sales* for highly diversified industries and *unit of production* for industries with one specified product/service (e.g., energy utilities)
Time period	Rather a long-term focus, as applied by official forecasts, such as the IEA or EIA (e.g., 2015)
Carbon usage t_0	Actual scope 1–2 GHG emissions
Carbon usage t_1	Estimated scope 1–2 GHG emissions that reflect company- and market-specific conditions
Carbon input prices t_0	Actual market prices for fossil fuels
Carbon output prices t_0	Actual costs for emission reduction certificates related to scope 1 GHG emissions; if an emission trading scheme is in place, application of the average price of allowances of year t_0 for all scope 2 GHG emissions
Carbon input prices t_1	Official market forecasts (e.g., IEA or EIA) for fossil fuel prices
Carbon output prices t_1	Average price of EU ETS allowances of year t_0 for all scope 1–2 GHG emissions in year t_1 in combination with an annual inflation rate of 2%

Note: IEA = International Energy Agency; EIA = Energy Information Agency; GHG = greenhouse gas; EU ETS = European Union Emission Trading Scheme.

the carbon performance within the time frame 2006–2015 for both companies. The price per ton of carbon for scope 1–2 carbon input has been calculated to be US$200 in 2006 and is assumed to increase to US$300 in 2015.[6] In 2006, neither company was subject to emission trading requirements, nor did either consider offsetting options; therefore, no carbon output prices have to be considered. The average EU ETS allowance price in 2006 was about US$25.4[7], which is applied for all scope 1–2 GHGs in 2015, taking into account an inflation rate of 2%.

Due to increasing stakeholder pressure, both companies undertake measures to optimize their carbon performance. Company A identifies areas for internal process improvements, which serves to hold fossil fuel consumption constant while production increases. Nevertheless, the company's efficiency for purchased energy does not improve and, thus, its energy consumption grows proportionately to the production increase. Company B takes advantage of governmental financial support programs and substitutes half of its fossil fuel consumption with renewable energy sources. Furthermore, the company increases

Company A

	2006	2015	2006	2015
Carbon usage	Carbon inputs		Carbon outputs	
Scope 1	18,000 t	18,000 t	66,000 t	66,000 t
Scope 2	12,500 t	19,392 t	45,833 t	71,103 t
Sum	30,500 t	37,392 t	111,833 t	137,103 t
Carbon costs				
Scope 1	3,600,000 $	5,400,000 $	0 $	1,998,721 $
Scope 2	2,500,000 $	5,817,481 $	0 $	2,153,244 $
Sum	6,100,000 $	11,217,481 $	0 $	4,151,965 $
Sales	90 Mio $	140 Mio $		
Carbon intensity	339 t/Mio $	268 t/Mio $	1,243 t/Mio $	982 t/Mio $
Carbon dependency	79 %		79 %	
	Carbon in- & outputs			
Carbon exposure	0.0678 $/$sales	0.1101 $/$sales		
Carbon risk	62 %			

Company B

	2006	2015	2006	2015
Carbon usage	Carbon inputs		Carbon outputs	
Scope 1	18,000 t	9,000 t	66,000 t	33,000 t
Scope 2	12,500 t	16,289 t	45,833 t	59,726 t
Sum	30,500 t	25,289 t	111,833 t	92,726 t
Carbon costs				
Scope 1	3,600,000 $	2,700,000 $	0 $	999,360 $
Scope 2	2,500,000 $	4,886,684 $	0 $	1,808,725 $
Sum	6,100,000 $	7,586,684 $	0 $	2,808,085 $
Sales	90 Mio $	140 Mio $		
Carbon intensity	339 t/Mio $	181 t/Mio $	1,243 t/Mio $	664 t/Mio $
Carbon dependency	53 %		53 %	
	Carbon in- & outputs			
Carbon exposure	0.0678 $/$sales	0.0745 $/$sales		
Carbon risk	10 %			

Figure 3 Carbon performance assessment of two companies. t = tons; Mio $ = million U.S. dollars.

operational energy efficiency and cuts down on scope 2 carbon inputs, although this is counterbalanced by a production-related increase in energy consumption.

Figure 3 shows the assessment results. For 2006, both companies have the same carbon output intensity of 1,243 tons of carbon per million U.S. dollars in sales. Due to its efforts, company A reduces its intensity, which results in a carbon dependency of 79%. Company B reduces its carbon intensity more strongly, bringing it down to 53%. This has implications for the monetary carbon performance: For 2006, both companies have a carbon exposure of 5.78 cents per dollar sales. Company B almost remains at this level and, thus, faces a moderate carbon risk of 10%. By contrast, company A has to cope with a significant increase in the financial relevance of carbon, expressed by a carbon risk of 62%. The increase in carbon costs results in an increase of its carbon risk, although the company reduces its carbon intensity. For policy makers, the assessment results indicate that the financial support program can be effective: Company B's carbon dependency is one third lower than company A's. For financial stakeholders, the assessment results illustrate that an investment in company B is less risky from a carbon perspective than investment in company A: Company B s likely increase of the financial relevance of carbon is six times lower.

Discussion and Conclusion

In this article, we discuss existing approaches for assessing the carbon performance of companies and conclude that there is no common agreement on how to conduct corresponding assessments. We suggest a holistic perspective on carbon flows on the micro level and define a systematic and comprehensive approach based on four corporate carbon performance indicators. We differentiate between the physical and monetary spheres and between current and future performance. The suggested specification of the indicators facilitates their practical use and aims at a standardized approach for analysis and reporting purposes.

We see two main stakeholders as users of the suggested indicators. For policy makers, the two physical indicators illustrate the carbon hot spots in a value chain that could be targeted for carbon reduction policies. The two monetary indicators reveal which carbon costs and risks lurk behind the business activities of companies and how companies' policies might affect their future competitiveness. As such, policy makers can utilize the indicators when formulating climate polices and as tools to evaluate whether existing policies have been effective. For financial markets, the physical indicators provide insights regarding a company's carbon management efforts and its optimization potentials. The monetary indicators show carbon's present and future financial implications for companies. Accordingly, information can be used to optimize investment portfolios or determine risk premiums. As such, actors in financial markets will be able to readjust their investment analyses and loan assessments and therefore help pave the way toward a low-carbon future. In addition to these two external stakeholders, companies can obtain insights regarding the carbon-related enhancements or risks of their production processes and facilities. On the basis of such information, they can assess future investments and projects or analyze existing processes. Moreover, they can utilize the corresponding information within marketing or corporate reporting.

Our suggestion for a standardized approach encompasses corporate scope 1–2 carbon usage. This raises the question of whether a life cycle–wide consideration would be more appropriate. Life cycle assessments (LCAs) can deliver precise results for the life cycle–wide carbon analysis of products and services.[3] If this is the purpose of an analysis, an LCA should be conducted. For the purpose of analyzing organizations (our aim in this article), the GHG protocol (WBCSD and WRI 2004) seems to be more appropriate. Because the applicability of scope 3 is rather complex, however, we suggest a specification based on a scope 1–2 approach. Nevertheless, future research should focus on facilitating a scope 1–3 approach. Most importantly, this requires clear industry-specific conventions on which carbon usage should be considered in scope 3, given the feasibility of data collection. Specialized service providers, such as Centre Info SA[9] or Trucost Plc,[10] can provide initial data sets as a basis for this research. Furthermore, organizations such as

FORUM

the Global Reporting Initiative[11] or the Carbon Disclosure Project[12] could extend their reporting frameworks to help gather these data. This also applies to other discussed data, such as carbon input costs or expenses relating to emissions trading. Such considerations could then further include the costs of internalizing external effects and could describe for stakeholders the overall negative carbon externalities caused by individual value chains.

Notes

1. See http://bie.berkeley.edu/files/Consumer FootprintCalc.swf.
2. In the Kyoto Protocol, six greenhouse gases are specified, which are measured according to their global warming potential in CO_2 equivalents: carbon dioxide, methane, nitrous oxide, hydrofluorocarbons, perfluorocarbons, and sulfur hexafluoride. To consider all six under the umbrella of *carbon* is consistent with the other approaches (e.g., the carbon footprint; Carbon Trust 2007). Furthermore, CO_2 accounts for the main proportion, as, on average, about 93% of GHG emissions of companies in the FTSE 100 (i.e., a share index of the 100 most highly capitalized companies listed on the London Stock Exchange) are CO_2-related (Henderson and Trucost 2005). For a description of the methods for calculating CO_2 equivalents, see IPCC 2007b.
3. Editor's note: For extensive analyses of ecoefficiency, see the special issue of this journal on ecoefficiency and industrial ecology, volume 9, number 4 (www3.interscience.wiley.com/journal/120129080/issue).
4. See www.ecoinvent.org; a list with further database providers can be found at http://lca.jrc.ec.europa.eu/lcainfohub/databaseList.vm.
5. See, for example, www.pointcarbon.com or www.wtrg.com.
6. To determine these costs, one has to consider separately the amount of each carbon input (for scope 1 and 2) and the related prices in t_0 and t_1. For example, carbon input prices in 2006 were \$719.30/t carbon for distillate fuel oil, \$474.52/t carbon for natural gas, and \$85.96/t carbon for other industrial coal (on the basis of our own calculations and EIA [2008, table A3]). Within this example, we assumed scope 1 and 2 to have the same average carbon costs in t_0 (\$200/t) and t_1 (\$300/t).
7. The average price was about EUR18.1 per ton, and we applied an exchange rate of US\$1.4/EUR. For simplicity, we disregarded the high volatility of this price in 2006.
8. This was one main outcome of the recent expert workshop 34th LCA Discussion Forum: Life Cycle Assessment Versus CO_2 Footprint? held on 14 March 2008, Lausanne, Switzerland (www.lcainfo.ch/DF/DF34/Program.htm).
9. See www.centreinfo.ch.
10. See www.trucost.com.
11. See www.globalreporting.org.
12. See www.cdproject.net.

References

Anderson, D. 2006. *Costs and finance of abating carbon emissions in the energy sector.* London: Imperial College London.

Ardenti, Y. M. and S. Gilardi. 2007. *The carbon intensity of car manufacturers.* Fribourg, Switzerland: Centre Info.

Bailey, E. 1998. *Intertemporal pricing of sulfur dioxide allowances.* Working paper, MIT Center for Energy and Environmental Policy Research, Cambridge, MA.

Barker, T., T. Foxon, J. Köhler, R. Gross, M. Leach, and P. Pearson. 2006. *Submission to the DTI energy review.* University of Cambridge, Imperial College London. www.berr.gov.uk/files/file30729.pdf.

Bartelmus, P., S. Moll, S. Bringezu, S. Nowak, and R. Bleischwitz. 2004. Translating sustainable development into practice: A "patchwork" of some new concepts and an introduction to material flows analysis. In *Eco-efficiency, regulation and sustainable business,* edited by R. Bleischwitz and P. Hennicke. Cheltenham, UK: Edward Elgar.

Bentley, R. W. 2002. Global oil & gas depletion: An overview. *Energy Policy* 30: 189–205.

Bosetti, V., C. Carraro, and M. Galeotti. 2006. The dynamics of carbon and energy intensity in a model of endogenous technical change. *Energy Journal* (Sp. Iss. 1): 191–205.

BP. 2006. *BP sustainability report 2006.* London: BP.

Brambles. 2006. *Annual report 2006.* Sydney.

Burritt, R. and C. Saka. 2005. Environmental management accounting applications and eco-efficiency: Case studies from Japan. *Journal of Cleaner Production* 14: 1262–1275.

Busch, T. and V. H. Hoffmann. 2007. Emerging carbon constraints for corporate risk management. *Ecological Economics* 62(3–4): 518–528.

Butner, K., D. Geuder, and J. Hittner. 2008. *Mastering carbon management—Balancing trade-offs to optimize supply chain efficiencies.* New York: IBM Institute for Business Value.

Campbell, C. J. 2005. *Oil crisis: Multi-science.* Brentwood, UK: Multi-Science Publishing.

Capoor, K. and P. Ambrosi. 2007. *State and trends of the carbon market 2007.* Washington, DC: World Bank.

Carbon Trust. 2006a. *Carbon footprints in the supply chain: The next step for business.* London: Carbon Trust.

Carbon Trust. 2006b. *Climate change and shareholder value.* London: Carbon Trust.

Carbon Trust. 2007. *Carbon footprinting: An introduction for organisations.* London: Carbon Trust.

CERES. 2006. *Global framework for climate risk disclosure—A statement of investor expectations for comprehensive corporate disclosure.* Boston: CERES.

Clarkson, R. and K. Deyes. 2002. *Estimating the social cost of carbon emissions.* London: Department of Environment, Food and Rural Affairs.

Deffeyes, K. S. 2003. *Hubbert's peak: The impending world oil shortage.* Princeton, NJ: Princeton University Press.

DeSimone, L. D. and F. Popoff. 1997. *Eco-efficiency, the business link to sustainable development.* Cambridge, MA: MIT Press.

E.on UK. 2006. *E.ON UK—changing energy—climate change: Our approach.* Coventry, UK: E.on UK.

EC (European Commission). 2007a. *Carbon footprint—what it is and how to measure it.* Brussels, Belgium: EC.

EC (European Commission). 2007b. *EU action against climate change.* Brussels, Belgium: EC.

Ehrenfeld, J. R. 2005. Eco-efficiency—philosophy, theory, and tools. *Journal of Industrial Ecology* 9(4): 6–8.

EIA (Energy Information Administration). 1995. *Measuring energy efficiency in the United States' economy: A beginning.* Washington, DC: EIA.

EIA. 2007. *Voluntary reporting of greenhouse gases program, fuel and energy source codes and emission coefficients.* Washington, DC: EIA.

EIA. 2008. *Annual energy outlook 2008.* Washington, DC: EIA.

EnCana. 2007. *Environmental performance.* Calgary, Canada: EnCana.

GRI (Global Reporting Initiative). 2006. *Sustainability reporting guidelines—environment performance indicators.* Amsterdam: GRI.

Henderson and Trucost. 2005. *The carbon 100—quantifying the carbon emissions, intensities and exposures of the FTSE 100.* London: Henderson Global Investors.

Hoffman, A. J. 2006. *Getting ahead of the curve: Corporate strategies that address climate change.* Philadelphia: Pew Charitable Trusts.

Holcim. 2005. *Corporate sustainable development report 2005.* St. Gallen, Switzerland: Holcim.

Horngren, C. T., G. L. Sundem, J. A. Elliott, and D. R. Philbrick. 2006. *Introduction to financial accounting,* 9th ed. Upper Saddle River, NJ: Pearson Prentice Hall.

Hourcade, J. C., D. Demailly, K. Neuhoff, and M. Sato. 2007. *Differentiation and dynamics of EU ETS industrial competitiveness impacts: Climate strategies report.* Cambridge: Climate Strategies.

Huesemann, M. H. 2006. Can advantages in science and technology prevent global warming? *Mitigation and Adaptation Strategies for Global Change* 11: 539–577.

Hutchinson, H. 2006. Climate change: When hell freezes over—European utilities. London: ING.

IEA (International Energy Agency). 2006. *World energy outlook 2006.* Paris: IEA.

Innovest. 2006. *Carbon disclosure project report 2006—global FT500.* London: Innovest Strategic Value Advisors.

IPCC (Intergovernmental Panel on Climate Change). 2005. *Carbon dioxide capture and storage.* New York: Cambridge University Press.

IPCC. 2007a. *Climate change 2007: The physical science basis—overview for policymakers.* Cambridge: Climate Strategies.

IPCC. 2007b. *Climate change: Synthesis report.* Geneva: IPCC.

Jones, C. A and D. L. Levy. 2007. North American business strategies towards climate change. *European Management Journal* 25(6): 428–440.

Kuik, O. 2003. Climate change policies, energy security and carbon dependency. *International Environmental Agreements: Politics, Law and Economics* 3: 221–242.

Lebel, L., P. Garden, M. E. N. Banaticla, R. D. Lasco, A. Contreras, A. P. Mitra, C. Sharma, H. T. Nguyen, G. L. Ooi, and A. Sari. 2007. Integrating carbon management into the development strategies of urbanizing regions in Asia—Implications of urban function, form and role. *Journal of Industrial Ecology* 11(2): 61–81.

Liberatore, A.. 2001. From Arrhenius to the Kyoto Protocol: Climate change and the interplay between science and policy. In *Knowledge, power and participation in environmental policy analysis,* edited by M. Hisschemöller et al. Edison, NJ: Transaction Publishers.

Lifset, R. 2007. Cement, yogurt, and mercury. *Journal of Industrial Ecology* 11(3): 1–3.

Llewellyn, J. 2007. *The business of climate change—challenges and opportunities.* New York: Lehman Brothers.

Maxime, D., M. Marcotte and Y. Arcand. 2006. Development of eco-efficiency indicators for the

Canadian food and beverage industry. *Journal of Cleaner Production* 14(14): 636–648.

Metz, B., O. Davidson, R. Swart, and J. Pan. 2001. *Climate change 2001: Mitigation.* Cambridge, UK: Cambridge University Press.

Montero, J. and A. Ellerman. 1998. *Explaining low sulfur dioxide allowance prices: The effect of expectation errors and irreversibility.* Technical document. Cambridge, MA: Center for Energy and Environmental Policy Research CEEPR, MIT.

Morioka, T., K. Tsunemi, Y. Yamamoto, H. Yabar, and N. Yoshida. 2005. Eco-efficiency of advanced loop-closing systems for vehicles and household appliances in Hyogo Eco-Town. *Journal of Industrial Ecology* 9(4): 205–221.

Nestlé. 2007. *Environment—key figures.* Vevey, Switzerland: Nestlé.

Nordhaus, W. D. 2006. *The Stern review on the economics of climate change.* NBER Working Paper No. W12741. Cambridge: National Bureau of Economic Research.

Peskin, H. M. and M. S. D. Angeles. 2001. Accounting for environmental services: Contrasting the SEEA and the ENRAP approaches. *Review of Income and Wealth* 47(2): 203–219.

Petro-Canada. 2007. *Climate change & greenhouse gases.* Calgary, Canada: Petro-Canada.

Raupach, M. R., G. Marland, P. Ciais, C. L. Quéré, J. G. Canadell, G. Klepper, and C. B. Field. 2007. Global and regional drivers of accelerating CO_2 emissions. *PNAS Early Edition* 1: 1–6.

Reynolds, D. B. 1999. The mineral economy: how prices and costs can falsely signal decreasing scarcity. *Ecological Economics* 31: 155–166.

Roche. 2007. *Roche's SHE performance—greenhouse gas emissions.* Basel.

Sabour, S. A. A. 2005. Quantifying the external cost of oil consumption within the context of sustainable development. *Energy Policy* 33(6): 809–813.

Schaltegger, S. and A. Sturm. 1990. Ökologische Rationalität: Ansatzpunkte zur Ausgestaltung von ökologisch-orientierten Managementinstrumenten. [Ecologic rationality: Starting points for the development of ecology-oriented management tools.] *Die Unternehmung* 4: 273–290.

Scholz, R. W. and A. Wiek. 2005. Operational eco-efficiency—comparing firms' environmental investments in different domains of operation. *Journal of Industrial Ecology* 9(4): 155–170.

Schultz, K. and P. Williamson. 2005. Gaining competitive advantage in a carbon-constrained world: Strategies for European business. *European Management Journal* 23(4): 383–391.

Societe Generale. 2007. *CREAM-ing carbon risk—European carbon winners and losers.* Paris: Societe Generale Cross Asset Research.

Stern, N. 2006. *The economics of climate change—the Stern review.* Cambridge, UK: HM Treasury.

Trexler, M. C. 2005. Of crystal balls and market fundamentals: Anticipating GHG prices. In *Green trading markets: Developing the second wave,* edited by P. C. Fusaro and M. Yuen. Amsterdam: Elsevier.

Toyota. 2006. *Sustainability report 2006.* Tokyo: Toyota.

Trucost. 2006. *Carbon counts—the Trucost carbon footprint ranking of UK investment funds.* London: Trucost Plc.

Urdal, B. T., M. Kopp, and T. Völker. 2006. *Carbonizing valuation—assessing corporate value at risk from carbon.* Zurich, Switzerland: SAM, WWF.

Vattenfall. 2007. *Climate map 2030.* Stockholm, Sweden: Vattenfall AB.

Voisin, S. and C. Lamotte. 2006. *Carbon impact on utilities.* Paris: Cheuvreux—Sustainable & Responsible Investment.

WBCSD (World Business Council for Sustainable Development). 2000. *Measuring eco-efficiency: A guide to reporting company performance.* Geneva, Switzerland: WBCSD.

WBCSD and WRI (World Resources Institute). 2004. *The greenhouse gas protocol: A corporate accounting and reporting standard* (revised version). Geneva, Switzerland: WBCSD.

Weidema, B. P., M. Thrane, P. Christensen, J. Schmidt, and S. Lokke. 2008. Carbon footprint: A catalyst for life cycle assessment? *Journal of Industrial Ecology* 12(1): 3–6.

Weinhofer, G. and V. H. Hoffmann. Forthcoming. Mitigating climate change—how do corporate strategies differ? *Business Strategy and the Environment.*

Weitzman, M. L. 1999. Pricing the limits to growth from minerals depletion. *Quarterly Journal of Economics* 114(2): 691–706.

Wieczorek, A. J., ed. 2003. *Carbon flows between eastern and western Europe (CFEWE)—final report.* Amsterdam: Institute for Environmental Studies Vrije Universiteit De Boelelaan 1087.

About the Authors

Volker Hoffmann is an assistant professor and **Timo Busch** is a Post-Doctor at the Swiss Federal Institute of Technology (ETH) in Zurich, Switzerland.

Annex

Paper III

Whither Carbonomics?

The Carbon Performance of the 100 largest US Electricity Producers

Timo Busch, Georg Weinhofer, Volker H. Hoffmann

ETH Zurich

2008, submitted to the *Journal of Industrial Ecology*

Abstract

Companies are likely to face a more carbon-constrained business environment in the future. On the one hand, tighter climate and energy policies may change the regulatory environment and, on the other hand, increases in the price of fossil fuels may significantly alter energy market conditions. Mainly two stakeholders are concerned with how these developments will be reflected in the carbon performance of companies: policy makers, as the performance illustrates the effectiveness of climate and energy policies; and financial stakeholders, as potential risks can be detected. To help meet their needs, this paper suggests an assessment framework and describes the effects of a more carbon-constrained business environment on electricity producers' carbon performance. While the US Energy Information Administration's (EIA) 'Annual Energy Outlook 2007 – Reference Case' is based on constant energy policy and market conditions, the US Senate bill 1766, the 'Low Carbon Economy Act of 2007,' assumes a more carbon-constrained business environment. Based on these two market forecasts and different corporate carbon strategies, we derive carbon scenarios and use them as the basis for an empirical carbon performance assessment of the 100 largest US electricity producers. This assessment places the emphasis on two dimensions, the physical carbon performance and the associated value in monetary terms. As one result we obtain a carbon risk ranking of the electricity producers.

Keywords

Carbon constraints, climate change regulation, fossil fuel prices, carbon risk ranking

Introduction

International climate and energy policy faces the challenge of decarbonizing the energy system (Nakicenovic 1997; Sun 2005), which will require significant changes from the status quo (see for example Sekar et al. 2007; Meier et al. 2005). This is especially true for countries such as the United States of America (US) where regulatory efforts towards a low-carbon society have been relatively limited compared to more progressive regions such as the European Union (EU) (Kolk and Hoffmann 2007; Bandyopadhyaya et al. 2007). As a result of changes in the business environment, such as regulatory efforts towards a low-carbon society and market pressures resulting from recent price increases for fossil fuels, carbon constraints are emerging that call into question traditional production and consumption patterns (Busch and Hoffmann 2007). Consequently, appropriate carbon management becomes increasingly relevant for companies from both a societal perspective – since they are responsible for a large share of anthropogenic greenhouse gas (GHG) emissions and must therefore play an important role in the decarbonization of the energy system – and from a firm perspective, because political as well as market-related developments are likely to affect the corporate financial bottom line.

Companies show different degrees of responsiveness to climate change issues, which is partly explained by their previous or declared corporate strategies (e.g., Kolk and Levy 2001; Hoffman 2006; Weinhofer and Hoffmann in press; Kolk and Pinkse 2005; Hoffmann 2007; Kolk and Pinkse 2004). Consequently, the emergence of carbon constraints and the prospects of a low-carbon future present different risks and opportunities to different companies. Research analyzing and comparing the carbon performance of companies (e.g., by measuring their GHG emissions) mostly pursues a static approach consisting of assessing the companies' current contribution to global warming (e.g., Ardenti and Flores 2007; Carbon Trust 2006; Henderson and Trucost 2005; Societe Generale 2007; Schultz and Williamson 2005).

However, only a dynamic view of corporate carbon performance can provide a holistic picture that combines firm characteristics with concrete emission scenarios. Such a dynamic view is especially important for external stakeholders: On the one hand, information about changes in carbon utilization and corporate emission patterns is important for policy makers when designing climate and energy policies. On the other hand, the magnitude of the financial consequences of such patterns and the changes in their carbon risk is of interest for financial markets.

Accordingly, this paper addresses the question of how external stakeholders can systematically assess the current and future carbon performance of companies. We build on four previously defined carbon performance indicators (Hoffmann and Busch in press) and derive a framework for assessing corporate carbon performance. We illustrate our methodology by focusing on the US electricity industry. Based on official US energy market forecasts released by the US Energy Information Administration (EIA) and on different corporate carbon strategies, we construct three carbon scenarios to estimate the carbon performance of the 100 largest US electricity producers. One of the outcomes of this is a carbon risk ranking of these companies. Such indicators can help provide policy makers with information on how they can accelerate improvements in the carbon performance of companies through low-carbon policies such as renewable energy subsidies (compare e.g. Jacobsson and Johnson 2000; Kydes 2007). Similarly, financial stakeholders[1] can utilize such information in order to assess the long-term profitability of their investments and decide whether it is prudent to maintain or increase investments in a given company.

The macro-micro link of carbon performance

Climate change has been acknowledged as a serious issue with a strong anthropogenic component and a corresponding need for human response in terms of adaptation and mitigation (IPCC 2007). For their part, policy makers are developing targets to curb GHG emissions and reduce carbon-based energy production. For example, the EU has set a greenhouse gas reduction target of 20% by 2020 and developed the vision to further reduce GHG by 60 to 80% by 2050 compared to 1990 (EC 2007). However, economies are limited in their ability to change their patterns of utilizing carbon inputs, a phenomenon termed carbon lock-in (Unruh 2000). Various policy approaches are conceivable to address such lock-in conditions, for example strategic niche management, which seeks to foster a market transformation towards low-carbon technologies (Unruh 2002). These and other approaches constitute important but only first steps towards decarbonizing the energy system. The ensuing steps have to focus on the emission sources and track whether the envisioned changes in fossil fuel consumption and corresponding GHG emissions occur and how further carbon reductions can be achieved. Therefore, policy makers require indicators that, on the one hand, incorporate different forecasts of future market conditions and policies and, on the other, show how macro level changes such as climate policies are reflected in activities on the micro level, e.g., by firms.

We argue that a similar situation obtains with respect to the financial dimension of climate change. Macro-analyses such as the Stern (2006) Report estimate the negative economic implications of global warming and quantify the costs of taking steps towards mitigation and adaptation. In response, the finance industry has put climate change and energy issues high on its agenda. This can be illustrated, for example, by the number of recently published industry reports on related topics (e.g., CERES 2002; Innovest 2007), the UNEP Financial Initiative,[2] or the Carbon Disclosure Project.[3] But for analyzing financial effects on their portfolios and anticipating the implications for future investments, financial analysts need indicators on the micro-level that take into account forecasts for future markets and price conditions and thereby accurately reflect the risks arising from carbon utilization for individual firms and for aggregated asset portfolios.

Several efforts in practitioner-oriented papers as well as in the scientific literature have been made that illustrate how changes in carbon utilization can be assessed on the micro level. For example, the World Business Council for Sustainable Development and the World Resources Institute developed the Greenhouse Gas Protocol (WBCSD and WRI 2004). Hoffman (2006) suggested metrics for measuring GHG emissions in order to track and report on emission reductions. Based on such metrics, Trucost (2006) developed a carbon ranking of investment funds. Other authors such as Ardenti & Flores (2007) and Voisin & Larrotte (2006) have used such information for the external analysis of companies. In addition, companies themselves such as BP (2006) and Nestlé (2007) employ such data for internal analysis and reporting purposes. Furthermore, some analyses go beyond the pure physical performance (i.e. carbon flows) and include monetary information (i.e. the costs of these carbon flows) in order to more holistically assess a company's carbon performance (Carbon Trust 2006; Henderson and Trucost 2005).

Two main shortcomings of previous comparative studies can be illustrated with respect to the physical and monetary dimension of the assessments. First, approaches that determine the physical carbon performance only consider the status quo of the companies and do not incorporate forecasts of future market developments. While this is sufficient for ex post analyses, assessments of the impacts of climate policies require an ex ante perspective. Furthermore, those studies that incorporate monetary information usually apply a hypothetical *future* carbon price to *current* carbon emissions to obtain a picture of the monetary carbon performance. However, in order to improve the quality of such assessments, changes in energy market conditions and their effects on future emissions should be taken into account. Also, besides risks from carbon outputs (i.e., GHG emissions), both carbon outputs *and* inputs should be included in the analysis. Accordingly, first the future physical performance has to be estimated based on market forecasts; afterwards, the future monetary performance can be obtained by combining this future physical performance with estimated future prices.

Framework for assessing corporate carbon performance

Based on this idea, Hoffmann and Busch (in press) have proposed a consistent set of four carbon performance indicators (Figure 1). First, carbon intensity measures a company's physical carbon flows and relates them to a business metric such as sales or amount of products. This static indicator illustrates the carbon performance in a given year from an environmental perspective. It can be calculated considering a company's current (t_0) carbon inputs (fossil fuels) or outputs (GHGs). Second, carbon dependency also focuses on physical flows but adds a scenario component by taking into account the extent to which the company will rely on carbon in the future (t_1). This dynamic view illustrates potential improvements in the carbon intensity over time (comparing t_1 with t_0) by including future market conditions and technology developments. Third, carbon exposure extends the analysis beyond the pure physical carbon flows to a monetary sphere and reveals the financial relevance of a company's carbon intensity. In order to incorporate all financial effects related to carbon, this indicator combines the carbon input and output dimension in one monetary figure for t_0. Fourth, carbon risk combines the insights regarding carbon dependency with monetary implications and allows the derivation of a company-specific carbon risk. Again, this last indicator takes a dynamic view and, therefore, supports estimates of the extent to which carbon will matter for a company from a financial perspective in the future, relative to today.

	Static approach	Dynamic approach
Physical units	Carbon Intensity in t_0 and t_1 focusing on fossil fuels (inputs) or alternatively CO_2 emissions (output) Input $\quad C_I In_{t_0} = \dfrac{\sum C_{I_{t_0}}}{BM_{t_0}} \qquad C_I In_{t_1} = \dfrac{\sum C_{I_{t_1}}}{BM_{t_1}}$ Output $\quad C_O In_{t_0} = \dfrac{\sum C_{O_{t_0}}}{BM_{t_0}} \qquad C_O In_{t_1} = \dfrac{\sum C_{O_{t_1}}}{BM_{t_1}}$	Carbon Dependency for timeframe $\Delta t = t_1 - t_0$ Input $\quad C_I De_{\Delta t} = \dfrac{C_I In_{t_1}}{C_I In_{t_0}} \cdot 100$ Output $\quad C_O De_{\Delta t} = \dfrac{C_O In_{t_1}}{C_O In_{t_0}} \cdot 100$
Monetary units	Carbon Exposure in t_0 and t_1 $CEx_{t_0} = \dfrac{\sum C_{i_{t_0}} \cdot p_{I_{t_0}} + \sum C_{O_{t_0}} \cdot p_{O_{t_0}}}{BM_{t_0}}$ $CEx_{t_1} = \dfrac{\sum C_{i_{t_1}} \cdot p_{I_{t_1}} + \sum C_{O_{t_1}} \cdot p_{O_{t_1}}}{BM_{t_1}}$	Carbon Risk for timeframe $\Delta t = t_1 - t_0$ $CRi_{\Delta t} = \left(\dfrac{CEx_{t_1}}{CEx_{t_0}} - 1\right) \cdot 100$

C_I Amount of carbon inputs (fossil fuels) p_I Price of fossil fuels t_0 Baseline year
C_O Amount of direct CO_2 emissions p_O Price of CO_2 emissions t_1 Target year
BM   Business metric (e.g., sales or product output)

Figure 1: Carbon Performance Indicators according to Hoffmann and Busch (in press)

Based on these four indicators, we suggest a carbon performance assessment framework for comparative analyses of companies.[4] It comprises three major steps (see Figure 2). Each of these steps consists of several sub-steps (boxes below the three major steps), which yield an interim assessment result that constitutes the basis for the eventual calculation of the four carbon performance indicators.

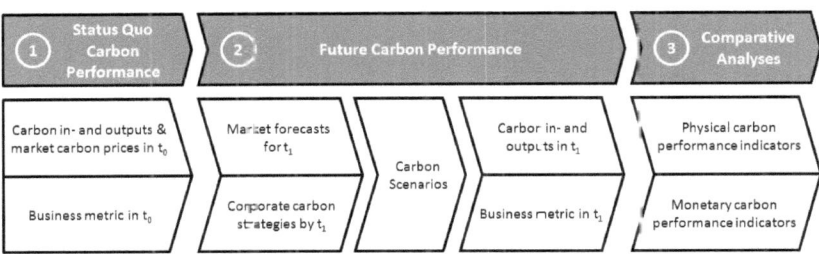

Figure 2: Corporate carbon performance assessment framework

In step one, data is collected to calculate each company's current (status quo) carbon performance. Their carbon inputs and outputs as well as market carbon prices have to be determined for a baseline year (t_0). Furthermore, a business metric as well as the corresponding value for each company have to be determined that represent the denominator for the static carbon performance indicators for t_0. In step two, data for the expected future carbon performance are collected. This step requires defining carbon scenarios. In our terminology, carbon scenarios describe how macro-level developments translate into micro-level strategies. As such, they combine market forecasts for carbon input-output patterns in an industry and corresponding prices for the target year (t_1) with possible corporate carbon strategies to be implemented by that year. Based on these carbon scenarios, the carbon inputs and outputs and business metric in t_1 have to be determined for each firm. In step three, a comparative analysis is conducted that evaluates the carbon performance of the companies in the different carbon scenarios. Thus, the four carbon performance indicators illustrated above (Figure 1) have to be determined based on the status quo (step one) and the estimated future (step two) carbon performance. In the following, we exemplify this framework in detail by applying it to the US electricity producing industry, an industry that is highly relevant for climate change mitigation and whose activities are based to a large extent on carbon (compare IEA 2006).

Data and assumptions

Using several publicly available data sources, we constructed a database consisting of data on the carbon inputs and outputs as well as electricity generated by the power plants owned by the 100 largest[5] US electricity producers for the timeframe 2004-2030. With the help of this database we then calculated these companies' carbon performance according to the framework outlined above. Seven of the 100 companies produced electricity solely from carbon-free energy resources and were therefore excluded from the further analysis.[6] In 2004 the resulting 93 companies produced 82% of the total electricity in the US and were responsible for 88% of the total CO_2 emissions from US electricity production (see Annex 1 for further details).

Carbon inputs and outputs and carbon prices for baseline year

For the baseline year of our analysis (2004), the company-specific carbon inputs are obtained through a combination of data on fossil fuel usage per power plant (EIA 2004b) and ownership shares per facility (CERES 2006). The carbon outputs per company are drawn from CERES' (2006) report on electricity producers' CO_2 emissions in 2004. Carbon input and output prices for the US in 2004 are based on the EIA's (2007) 'Annual Energy Outlook 2007' (see Table 1).

Business metric in baseline year

We use the amount of generated electricity per year as the business metric. Similarly to the carbon inputs, data on electricity production in 2004 for each of the companies is obtained from a combination of net electricity output per power plant (EIA 2004b) and ownership shares per facility (CERES 2006).

Market forecasts for target year

Two official market forecasts for the year 2030 form the basis of our analysis. They contain all relevant information on the electricity market such as production capacity, electricity generation, fossil fuel prices, and energy mix. The first is based on EIA's (2007) 'Reference Case,' which anticipates no significant changes in climate policies or in carbon and electricity market conditions for 2030 (see Table 1). The second forecast is based on the US Senate bill 1766, the 'Low Carbon Economy Act of 2007' (LOC 2007). This bill prescribes a mandatory CO_2 emission allowance program in the US that would result in a more carbon-constrained business environment.[7] Based on this bill, the EIA (2008) calculated corresponding data regarding the 2030 energy market conditions (notably 25 US$ per tonne of CO_2 emissions), which we use as a second forecast. As each of the forecasts is drawn from one source only, our analysis has the advantage that the underlying energy-economic models are internally consistent, i.e. the interactive effects between fossil fuel prices, generation capacity, and market demand and supply are incorporated. In order to exclude effects of inflation, we consider 2030 prices in real terms (in 2005 US$; see Table 1).

		2004	2030	
			EIA Annual Energy Outlook 2007 Reference Case	EIA Analysis of S.1766 Limited Alternatives Case
		(EIA 2007)	(EIA 2007)	(EIA 2008)
US Electricity Production and corresponding CO_2 Emissions				
Coal	[TWh]	1,979	3,330	2,505
Petroleum	[TWh]	122	107	67
Natural Gas	[TWh]	710	937	1,383
Nuclear Power	[TWh]	789	896	896
Renewable Sources	[TWh]	361	519	671
Others	[TWh]	12	9	0
Total	[TWh]	3,973	5,798	5,522
Total CO_2 Emissions	[million tonnes]	2,309	3,338	2,803
US Power Plant Retirements by 2030				
Coal	[GW]		5.5	11.1
Petroleum	[GW]		15.5	23.6
Natural Gas	[GW]		38	65.3
Nuclear Power	[GW]		2.9	2.9
Renewable Sources	[GW]		0	0
Total	[GW]		61.9	102.9
US Carbon In- and Output Prices				
Coal	(2005 US$ per short ton)	28.1	33.5	78.6
Petroleum (DFO)	(2005 US$ per barrel)	55.5	62.8	75.3
Petroleum (RFO)	(2005 US$ per barrel)	31.4	43.1	79.9
Natural Gas	(2005 US$ per mega cubic feet)	6.3	6.5	8.5
CO_2	(2005 US$ per metric ton)	0	0	25.0

Table 1: Data on the US energy market in 2004 and 2030

Corporate carbon strategies by target year

In the electricity-producing industry, future carbon inputs and outputs are tied to investments in new power plants. These investments constitute a particularly proprietary part of corporate strategies on which detailed public information is scarce. Therefore, we differentiate between two prototypical carbon strategies companies might pursue by 2030: a strategy that displays path-dependent[8] carbon utilization as well as a low-carbon strategy. The former assumes that the companies will rely on the energy sources they used in 2004 and therefore invest only in those sources through 2030. The latter reflects the option of reducing carbon utilization to its potential minimum and assumes that companies make new investments only in renewable energy sources. Although it is unlikely that all companies would simultaneously pursue only the low-carbon strategy, the strategy illustrates a plausible range of the carbon reduction potential for each company. Table 3 illustrates the detailed assumptions made for both strategies.

Carbon scenarios

In order to assess the electricity producers' carbon performance, we define three carbon scenarios by combining the market forecasts and the corporate carbon strategies (see Table 2): a 'business as usual' scenario (BAU), a 'carbon constraints' scenario (CACO), and a 'renewable energy' scenario (RENEW). The BAU and CACO scenarios are based on different market forecasts but consider no change in corporate carbon strategy (i.e., path-dependent behavior) and, therefore, allow an evaluation of the effects of different future energy market conditions. In contrast, the CACO and RENEW scenarios are based on the same market forecast ('Limited Alternatives Case'), but consider different corporate carbon strategies and, thus, illustrate the effects of adjusting the carbon strategy towards an accelerated use of renewable energy sources.[9]

		Market Forecast	
		EIA Annual Energy Outlook 2007 Reference Case	EIA Analysis of S.1766 Limited Alternatives Case
Corporate Carbon Strategy	Path-Dependent	'Business as Usual' (BAU)	'Carbon Constraints' (CACO)
	Low-Carbon		'Renewable Energy' (RENEW)

Table 2: Analyzed Carbon Scenarios

Carbon inputs & outputs in target year

In order to estimate the company-specific carbon inputs and outputs in 2030 in each of the carbon scenarios, we total the corresponding data for remaining, replaced, and additional electricity generation in that year. We calculate the remaining generation by taking electricity production from all power plants operational in 2004 and subtracting generation from retiring power plants. To estimate which plants retire, we assume that the oldest operating power plants (EIA 2004a) retire first, up to the retiring capacities reported in the underlying market forecast (Table 1). We then assign this retiring generation to individual companies based on ownership shares per facility (CERES 2006). The 2030 carbon inputs and outputs stemming from replaced and additional generation differ according to the underlying carbon strategies and are explained in Table 3.

		Replaced Generation	Additional Generation	2030 Generation
Corporate Carbon Strategy	Path-Dependent	Retired generation is replaced with its previous energy source but is produced with improved (average) efficiency of new US power plants for the respective energy source by 2030	Additional generation is produced with corresponding energy sources and average efficiencies of new US power plants for the respective energy source by 2030	Each company has same share of total US electricity generation per energy source in 2030 as in 2004
	Low-Carbon	Retired generation is replaced by renewable energy sources	Additional generation is produced by renewable energy sources	Each company has same total generation as in path-dependent carbon strategy

Table 3: Assumptions for corporate carbon strategies by 2030

Business metric in target year

The amount of electricity produced by each of the companies in 2030 is determined by the assumptions made for the three carbon scenarios regarding the US electricity production in 2030 in the market forecasts (Table 1) and the corporate carbon strategies (Table 3). This step completes the information-gathering phase. With that, the comparative analysis of the companies' carbon performance can be conducted.

Results

Physical carbon performance indicators

Figure 3 illustrates the results for the carbon intensities for the 93 companies in 2004 and 2030 based on the underlying assumptions of the three carbon scenarios.[10] Carbon dependency, the ratio of future and status quo carbon intensity, increases counterclockwise from the horizontal axis (corresponding to a value of 0%) to the vertical axis, while diagonal lines show an equal carbon dependency.

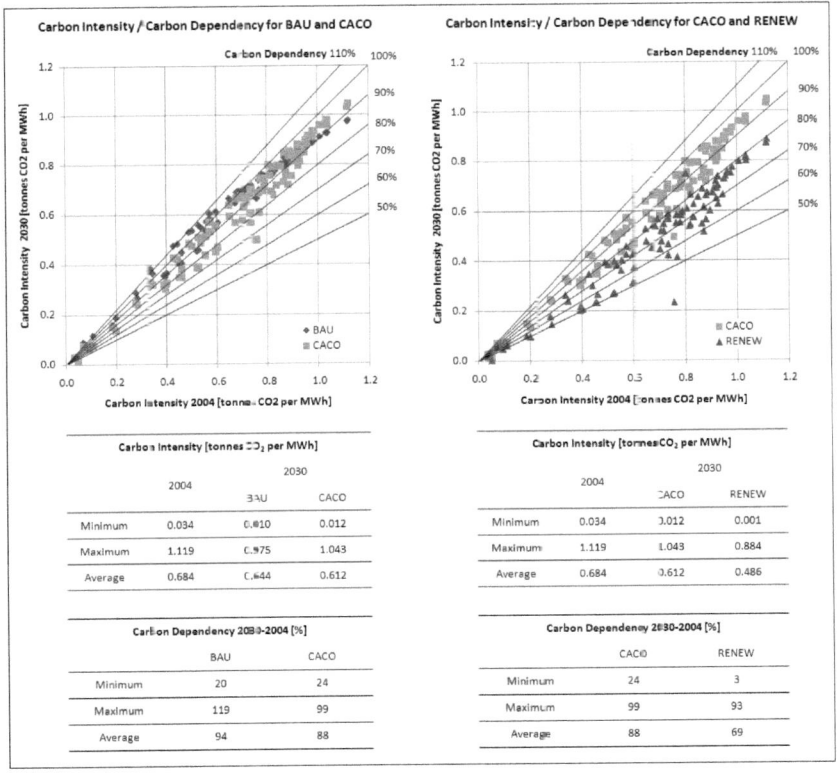

Figure 3: Physical carbon performance

As an important first result, there is a big spread between the carbon intensities (left part of Figure 3): in 2004, the company at the upper end has a carbon intensity of about 1.1 tonnes CO_2 per MWh, which is about 33 times higher than the carbon intensity of the company at the lower end of the spectrum. This spread increases to nearly a factor of 100 in the BAU and CACO scenarios because both scenarios assume a path-dependent investment behavior. The comparison between the two scenarios illustrates that the average carbon intensity in 2030 in the BAU scenario is just slightly lower compared to 2004 and again only slightly lower in the CACO scenario. Accordingly, the average carbon dependency is on a rather high level at 94%

(BAU) and 88% (CACO). For the CACO scenario this is surprising because a major underlying assumption of this scenario is that the business environment is much more carbon-constrained, which should result in less carbon-intense production patterns.

A comparison of the carbon scenarios CACO and RENEW (right part of Figure 3) illustrates that for all companies the change from a path-dependent to a low-carbon strategy results in a reduction of carbon intensity. With around 0.5 tonnes CO_2 per MWh, the average carbon intensity in the RENEW scenario is more than 20% lower compared to the CACO scenario and 25% lower than in the BAU scenario. Similar results are obtained for carbon dependency: on average companies are able to significantly reduce their carbon dependency. To summarize, following a low-carbon strategy results in a much lower carbon intensity and carbon dependency compared to the two scenarios that assume a path-dependent carbon strategy.

Monetary carbon performance indicators

Turning to the monetary carbon performance of the electricity producers, Figure 4 illustrates the results for the carbon exposures of the 93 companies in 2004 and 2030. Carbon risk relates future and current carbon exposure and increases counterclockwise from the horizontal axis to the vertical axis while diagonal lines show an equal carbon risk.

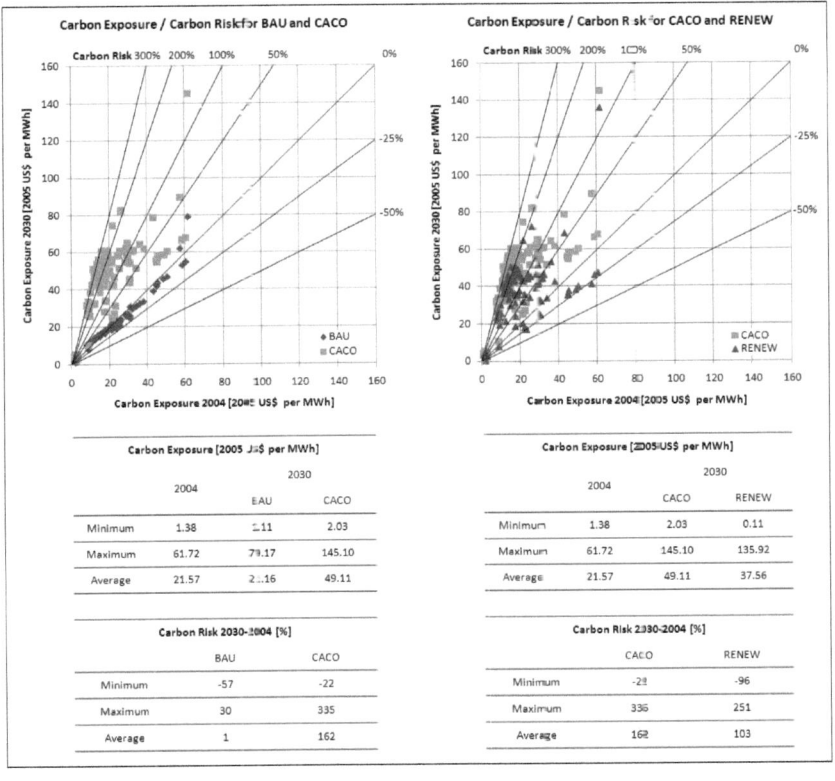

Figure 4: Monetary carbon performance

As one key result of taking into account monetary effects, the carbon performance picture becomes much more heterogeneous compared to an analysis based only on physical carbon flows. While the average carbon exposure in the BAU scenario remains at the 2004 level of about 21 US$ per MWh, it significantly increases in the CACO scenario to almost 50 US$ per MWh (left part of Figure 4). Companies at the lower end reduce their carbon risk in both scenarios, while at the upper end the carbon risk reaches a maximum of 30% in the BAU scenario and 335% in the CACO scenario. The latter case thus entails more than a tripling of

17

the financial relevance of carbon. Furthermore, in the BAU scenario the average carbon risk is very low at 1%, with a comparatively small variation around the mean. Hence, on average companies do not have to face negative financial implications from their carbon utilization. However, this picture changes significantly when looking at the CACO scenario with an average carbon risk of 162% and much larger variations.

A comparison of the CACO and RENEW scenarios (right part of Figure 4) shows that, similarly to the results obtained for the physical carbon performance, all companies are able to reduce their carbon exposure through a switch to the low-carbon strategy. As a general trend, the change in carbon strategy allows companies to partly reverse the risk increase that stems from the difference between the two market forecasts. However considering the average risk, this reversal is not substantial. Rather, the drastic increase in carbon prices in the underlying market forecast is so dominant that, on average, even if companies focus purely on investing in renewable energy sources, they still face an increase in carbon risk in the RENEW scenario.

However, especially for financial stakeholders, an important question is whether the difference in carbon risk is constant across the scenarios, i.e. if all companies face an equal average risk, or if any characteristics of individual companies can be detected. In order to shed light on this question, Figure 5 illustrates the company-specific carbon risks in the three carbon scenarios sorted by the BAU scenario results.

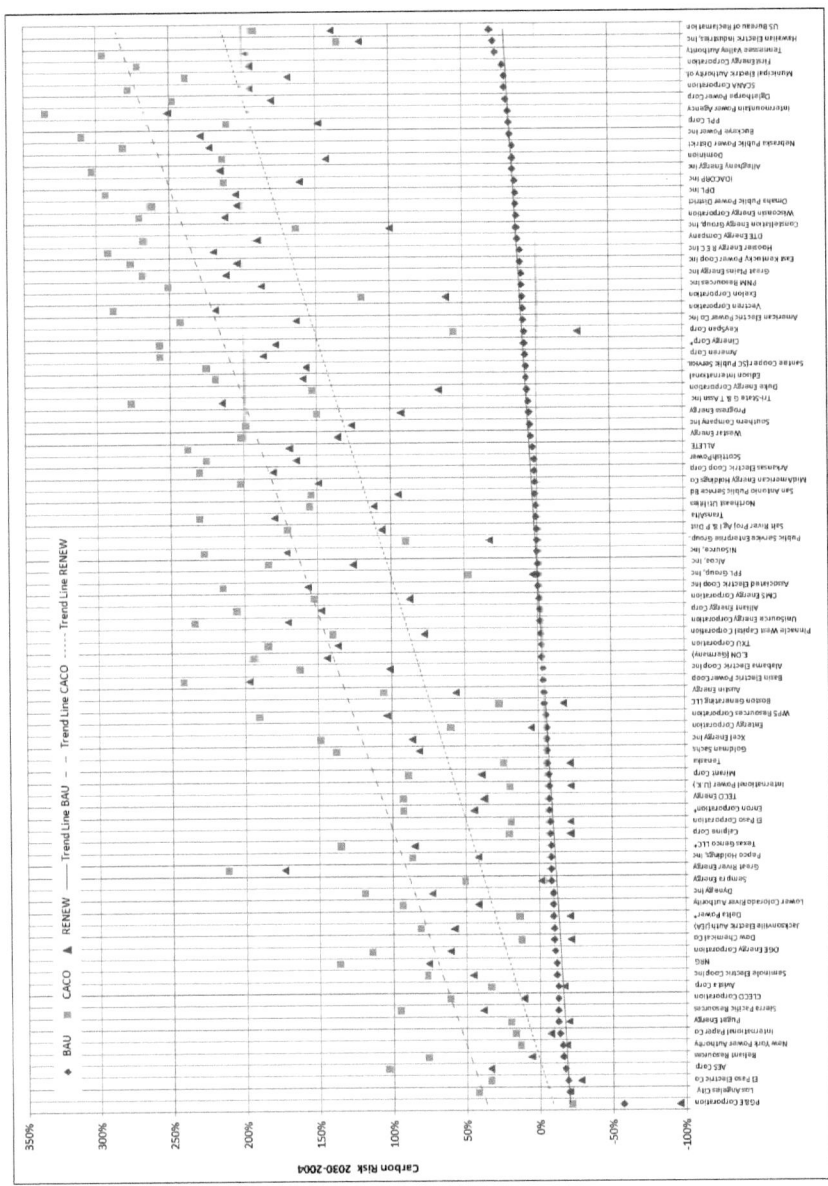

Figure 5: Carbon Risk: EAU – CACO - RENEW (sorted by increasing carbon risk in BAU)

The different slope of the trend lines of the BAU and CACO scenarios implies that in general, companies with a high carbon risk in the BAU scenario face an even relatively higher carbon risk in the CACO scenario. Hence, companies that do have a high carbon risk based on conservative assumptions will have an even higher risk in a more carbon constrained business environment. However, there is a large variation around the trend line of the CACO scenario. In contrast, the analogous consideration of the CACO and RENEW scenarios reveals that the slopes of the trend lines in Figure 5 are almost equal. Consequently, the potential to reduce the carbon risk is on average almost the same when switching from a path-dependent to a low-carbon strategy, irrespective of whether the company has a high or low carbon risk. In turn, this implies that companies with a high risk have a lower relative risk reduction potential when investing in renewable energy sources. Nevertheless, a significant variation around the trend line of the RENEW scenario can be observed, stressing the importance of detailed company-specific assessments.

Discussion

Interpreting our results from a policy perspective shows that there is a rather marginal change in the physical carbon performance of the US electricity producers when assuming a more carbon-constrained business environment compared to the business as usual case. This can be ascribed to the scenarios' assumptions regarding the US energy market in 2030: First, the lower electricity production from coal in the CACO scenario compared to the BAU scenario is partly compensated by the higher generation from natural gas, thereby substituting one fossil resource for another. Second, the prediction for the total US electricity production in the CACO scenario is lower than in the BAU scenario. As carbon intensity is measured in relative terms, the absolute carbon reduction due to changing production from coal to gas is partially compensated for by the lower total electricity generation. This is a very important outcome for policy makers: even though the total CO_2 emissions of the 100 electricity producers decrease from 2,956 in the BAU scenario to 2,429 million tonnes CO_2 in the CACO scenario (see Annex 1), the effect in terms of carbon intensity is almost negligible. Accordingly, policies inducing a more carbon-constrained business environment may be effective in reducing total emissions but may lead to only marginal enhancements in terms of carbon dependency of the US electricity mix.

Considering these results from the perspective of a financial stakeholder, an important question is whether the marginal change in physical carbon performance also implies a negligible change in monetary carbon performance. Interestingly, this is in fact the case in the BAU scenario, as there the average carbon risk is almost zero. Due to the underlying assumptions in this scenario, on average companies do not have to face negative financial consequences from their carbon utilization. However, these assumptions are rather conservative and do not incorporate any change towards a more carbon-constrained business environment, quite an unlikely eventuality in light of the urgency of the problem and current developments in international climate policy. By contrast, in the CACO scenario the costs for carbon inputs and outputs will increase significantly for most of the firms. As a consequence, companies face a carbon risk of 152% on average. These results stress the importance for financial markets to systematically integrate developments in carbon markets and the effects of climate policy into standard financial assessments in order to optimize the risks of their assets with respect to emerging carbon constraints. Notably, the high variation of the risk (from -22% to 335%) illustrates the different risk propositions of individual companies. Beyond that, it is important for policy makers as well as financial stakeholders to consider company-specific carbon strategies. In the RENEW scenario, the average carbon dependency is about 69%. Considering the average carbon dependency of 94% in the BAU scenario and

88% in the CACO scenario, we conclude that a change in carbon strategy is able to contribute three quarters of the overall reduction in carbon intensity. This emphasizes the relevance for policy makers of actively supporting investments in renewable energy sources because they appear to be more effective at reducing carbon intensities of individual firms than purely focusing on constraining carbon usage. From a financial point of view, the results in the RENEW scenario illustrate possible first mover advantages, as the overall potential to switch to renewable energy sources is limited. This argument is bolstered when considering the results of the carbon risk ranking (see Annex 2): The key features of high-ranked (i.e., low-risk) companies are a high share of power generation from renewable sources and/or a high share of gas within their fossil fuel electricity production. In contrast, the low-ranked companies can be characterized as having no electricity production from carbon-free energy sources and relying on coal and oil as their primary energy source. Therefore, financial stakeholders should particularly monitor the investment plans of companies with a high carbon risk and analyze whether they are able to partly reduce the carbon risks induced by the underlying market forecasts by investing in renewable energy sources.

Conclusion

Energy policy has to take into account corporate contributions to global warming when determining and evaluating climate regulation, but there is no internationally standardized accounting approach for doing so. Financial markets have just started to considerer the carbon performance of firms they have a stake in, but they are still struggling to find an appropriate method. This paper has contributed to this discussion by suggesting a corporate carbon performance assessment framework and providing the results of applying this framework to the US electricity sector. Furthermore, it sheds light on the differential effects of policy measures that constrain the carbon market as opposed to those that actively stimulate the accelerated use of renewable energy sources.

According to the EIA's 'Annual Energy Outlook - Reference Case,' US electricity producers should not face severe carbon risks. But the underlying predictions are very optimistic; instead, the institutional environment and the US energy market conditions are likely to become more carbon-constrained. Taking this view, we have developed a carbon risk ranking (based on the carbon risks of the companies in the CACO scenario, see Annex 2). If a path-dependent carbon strategy is assumed for the companies, our results provide a heterogeneous picture: top-ranked companies will approximately maintain their level of carbon exposure in a carbon-constrained business environment, whereas for the last-ranked company the financial significant from using and emitting carbon will approximately triple. We therefore emphasize the importance of financial stakeholders' consideration of the individual carbon strategies of companies.

Future research could quantitatively investigate whether these indicators already show statistically significant impacts on the financial performance of companies. Qualitative investigations could build on our approach and identify and analyze interesting cases in detail. Most importantly, such investigations could put more emphasis on the specific business environment and response strategies of individual companies. Such research would provide further important insights into the carbon world for both policy makers and financial markets.

Notes

[1] As financial stakeholders we include a company's owners such as shareholders (in the case of publicly traded corporations) or local communities (for companies not traded). Furthermore, we also include financial institutions within this term, e.g. lending banks with a stake in the companies.

[2] See http://www.unepfi.org

[3] See http://www.cdproject.net

[4] As the purpose of this paper is to conduct a comparative analysis across companies, the suggested assessment framework is focused on collecting data for more than one company. Nevertheless, each step can also be conducted individually for a single company. Corresponding results over several years can be utilized for a company's internal carbon performance controlling.

[5] As measured by electricity production in 2004.

[6] Center Point Energy, Energy Northwest, North Carolina Municipal Power Agency, PUD of Chelan County, PUD of Grant County, Seattle City Light, and US Corps of Engineers

[7] This mandatory CO_2 emission allowance program assumes that every year regulated companies have to submit the number of allowances equal to their CO_2 emissions to the US government. While companies get a certain number of allowances allocated for free (grandfathering), the remaining number of necessary allowances have to be bought on the market. The number of free allowances for companies results from the allocation process stipulated in the US Senate bill 1766. Accordingly, the US government issues a certain number of allowances each year (4,819 million in 2030). These allowances are allocated differently each year: in 2030 53% by auction, 25% for free to industries, 13% for free to set-aside emission programs, and 9% for free to states. Electricity producers get 54% of the 25% free allowances allocated to companies in 2030, namely 651 million allowances. The individual electricity producers' proportion of the freely allocated allowances is equal to their share in accumulated CO_2 emissions of all US electricity producers (LOC, 2007).

[8] The term 'path-dependent' usually refers to the conditions under which a company's current and future decisions depend on previous decisions and the patterns according to which these decisions were made (see for example Cowan and Gunby 1996). In our context this entails that in the future each company continues to operate in the same way as before, i.e. the company's strategy is to pursue the same path regarding the utilization of carbon-based technologies and related GHG emissions.

[9] We do not consider a carbon scenario based on the low-carbon strategy and EIA's AEO2007 Reference Case market forecast since an accelerated use of renewable energy sources appears very unlikely under the market conditions prevailing in the reference case.

[10] Only the output dimension is considered for carbon intensity, as there is no relative difference between the carbon input and output intensity. This is due to the fact that all utilized carbon inputs result in CO_2 emissions, since the underlying market forecasts do not assume the operation of Carbon Capture and Storage.

References

Ardenti, Y. M. and R. Flores. 2007. *The Carbon Intensity of Car Manufacturers - An updated sector study using envIMPACT®, the carbon risk analysis tool for fund managers.* Fribourg: Centre Info.

Bandyopadhyaya, G., F. Bagherib, and M. Mann. 2007. Reduction of fossil fuel emissions in the USA: A holistic approach towards policy formulation. *Energy Policy* 35(2): 950-965.

BP. 2006. *BP Sustainability Report 2006.*

Busch, T. and V. H. Hoffmann. 2007. Emerging carbon constraints for corporate risk management. *Ecological Economics* 62(3-4): 518-528.

Carbon Trust. 2006. *Climate change and shareholder value.* London: Carbon Trust.

CERES. 2002. *Value at Risk: Climate Change and the Future of Governance.* Boston: CERES.

CERES. 2006. *Benchmarking Air Emissions of the 100 largest Electric Power Producers in the United States – 2004.* Boston, MA: CERES.

Cowan, R. and P. Gunby. 1996. Sprayed to death: Path dependence, lock-in and pest control strategies *Economic Journal* 106(436): 521-542.

EC. 2007. *EU action against climate change.* Brussels: European Commission.

EIA. 2004a. *Annual Electric Generator Report* F860Y04. Washington, DC: Energy Information Administration.

EIA. 2004b. *Power Plant Database: Monthly Generation and Fuel Consumption.* F906920Y04. Washington, DC: Energy Information Administration.

EIA. 2007. *Annual Energy Outlook 2007.* Washington, DC: Energy Information Administration.

EIA. 2008. *Energy Market and Economic Impacts of S. 1766, the Low Carbon Economy Act of 2007.* Washington, DC: Energy Information Administration.

Henderson and Trucost. 2005. *The Carbon 100 - Quantifying the Carbon Emissions, Intensities and Exposures of the FTSE 100.* London: Henderson Global Investors.

Hoffman, A. J. 2006. *Getting Ahead of the Curve: Corporate Strategies that Address Climate Change.* Michigan: Few Trusts.

Hoffmann, V. H. 2007. EU ETS and investment decisions: the case of the German electricity industry. *European Management Journal* 25(6): 464-474.

Hoffmann, V. H. and T. Busch. in press. Corporate Carbon Performance Indicators: Carbon Intensity, Dependency, Exposure, and Risk. *Journal of Industrial Ecology.*

IEA. 2006. *CO_2 Emissions from Fuel Combustion 1971-2004.* Paris: International Energy Agency.

Innovest. 2007. *Carbon Beta and Equity Performance - An Empirical Analysis*. New York: Innovest Strategic Value Advisors.

IPCC. 2007. *Climate Change 2007: The Physical Science Basis*. Edited by Working Group I Contribution to the Fourth Assessment Report of the IPCC Intergovernmental Panel on Climate Change, *Intergovernmental Panel on Climate Change*. Cambridge: Cambridge University Press.

Jacobsson, S. and A. Johnson. 2000. The diffusion of renewable energy technology: an analytical framework and key issues for research. *Energy Policy* 28(9): 625–640.

Kolk, A. and D. Levy. 2001. Winds of Change: Corporate Strategy, Climate Change and Oil Multinationals. *European Management Journal* 19(5): 501-509.

Kolk, A. and J. Pinkse. 2004. Market Strategies for Climate Change. *European Management Journal* 22(3): 304-314.

Kolk, A. and J. Pinkse. 2005. Business responses to climate change: Identifying emergent strategies. *California Management Review* 47(3): 6-20.

Kolk, A. and V. H. Hoffmann. 2007. Business, climate change and emissions trading: Taking stock and looking ahead. *European Management Journal* 25(6): 411-414.

Kydes, A. S. 2007. Impacts of a renewable portfolio generation standard on US energy markets. *Energy Policy* 35(2): 809-814.

LOC. 2007. *US Senate Bill 1766*. Washington, DC: The Library of Congress.

Meier, P. J., P. P. H. Wilson, G. L. Kulcinski, and P. L. Denholm. 2005. US electric industry response to carbon constraint: a life-cycle assessment of supply side alternatives. *Energy Policy* 33(9): 1099-1108.

Nakicenovic, N. 1997. Freeing Energy from Carbon. In *Technological Trajectories and the Human Environment*, edited by J. H. Ausubel and H. D. Langford. Washington, D.C.: National Academy Press.

Nestlé. 2007. *Environment - Key Figures*.

Schultz, K. and P. Williamson. 2005. Gaining Competitive Advantage in a Carbon-constrained World: Strategies for European Business. *European Management Journal* 23(4): 383-391.

Sekar, R. C., J. E. Parsons, H. J. Herzog, and H. D. Jacoby. 2007. Future carbon regulations and current investments in alternative coal-fired power plant technologies. *Energy Policy* 35(2): 1064-1074.

Societe Generale. 2007. *CREAM-ing carbon risk - European carbon winners and losers*. Paris: Societe Generale Cross Asset Research.

Stern, N. 2006. *The Economics of Climate Change – The Stern Review*. Cambridge: Cambridge University Press.

Sun, J. W. 2005. The decrease of CO2 emission intensity is decarbonization at national and global levels. *Energy Policy* 33(8): 975-978.

Trucost. 2006. *Carbon Counts - The Trucost Carbon Footprint Ranking of UK Investment Funds*. London: Trucost Plc.

Unruh, G. C. 2000. Understanding carbon lock-in. *Energy Policy* 28(12): 817-830.

Unruh, G. C. 2002. Escaping carbon lock-in. *Energy Policy* 30(4): 317-325.

Voisin, S. and C. Lamotte. 2006. *Carbon Impact on Utilities*. Paris: Cheuvreux - Sustainable & Responsible Investment.

WBCSD and WRI. 2004. *The Greenhouse Gas Protocol: A Corporate Accounting and Reporting Standard (Revised version)*. Geneva, Washington DC: World Business Council for Sustainable Development, World Resources Institute.

Weinhofer, G. and V. H. Hoffmann. in press. Mitigating Climate Change - How do Corporate Strategies Differ? *Business Strategy and the Environment*.

Annex 1: Information on the analyzed electricity producers

Annex 1 illustrates the cumulative electricity production per fuel type and CO_2 emissions of the 93 electricity producers

- based on our calculations with data from CERES (2006) and EIA (2007; 2008)
- % values in 2004 represent the share of total US power generation per energy source and of total CO_2 emissions from electricity production

		2004	2030		
			BAU	CACO	RENEW
Coal	[TWh]	1,759 (89%)	2,960	2,227	1,707
Petroleum	[TWh]	105 (86%)	92	58	51
Natural Gas	[TWh]	482 (68%)	636	938	424
Nuclear	[TWh]	745 (94%)	846	846	724
Renewable	[TWh]	156 (43%)	224	289	1,452
Total	[TWh]	3,247 (82%)	4,771	4,358	4,358
CO2 emissions	[million tonnes]	2,027 (88%)	2,956	2,429	1,905

Annex 2: Carbon Risk Ranking

Rank based on Carbon Risk in CACO	Company Name	Electricity Generation (TWh) 2004	Electricity Generation Portfolio 2004					Carbon Intensity (tonnes CO₂ per MWh) 2030				Carbon Dependency 2030-2004				Carbon Exposure (2005 US$ per MWh) 2004				2030			Carbon Risk 2030-2004		
			COAL	OIL	GAS	NUC	RENEW	2004	BAU	CACO	RENEW	BAU	CACO	RENEW	2004	BAU	CACO	RENEW	BAU	CACO	RENEW	BAU	CACO	RENEW
1	PG&E Corporation	26	0%	0%	3%	58%	39%	0.051	0.010	0.012	0.001	20%	24%	3%	2.59	1.11	2.03	0.11				-57%	-22%	-96%
2	Dow Chemical Co	8	0%	0%	100%	0%	0%	0.527	0.457	0.386	0.270	87%	73%	51%	60.49	54.53	67.90	47.42				-10%	12%	-22%
3	New York Power Authority	24	0%	3%	13%	0%	84%	0.095	0.072	0.062	0.045	76%	65%	48%	9.22	7.83	10.45	7.54				-15%	13%	-18%
4	Delta Power*	13	0%	0%	99%	0%	0%	0.521	0.453	0.383	0.268	87%	73%	51%	50.57	37.58	68.47	46.14				-10%	35%	-21%
5	International Paper Co	9	11%	8%	19%	0%	61%	0.190	0.188	0.121	0.097	93%	60%	43%	21.48	20.71	23.10	19.96				-14%	17%	-7%
6	El Paso Corporation	8	0%	0%	100%	0%	0%	0.452	0.400	0.348	0.232	89%	77%	51%	50.65	46.81	60.31	39.85				-8%	19%	-21%
7	Puget Energy	2	0%	0%	43%	0%	57%	0.185	0.157	0.148	0.098	85%	80%	53%	22.70	19.86	27.14	18.11				-13%	20%	-20%
8	International Power (U.K.)	10	0%	0%	100%	0%	0%	0.403	0.363	0.322	0.207	90%	80%	51%	49.00	45.51	58.66	38.20				-7%	20%	-22%
9	Calpine Corp	85	0%	0%	93%	0%	7%	0.195	0.150	0.311	0.203	89%	84%	51%	45.44	41.82	54.71	35.60				-8%	20%	-22%
10	Tenaska	11	0%	0%	100%	0%	0%	0.391	0.354	0.316	0.201	96%	81%	51%	44.98	42.33	55.74	35.27				-6%	24%	-22%
11	Boston Generating LLC	11	0%	0%	93%	0%	0%	0.459	0.406	0.346	0.233	88%	75%	51%	45.58	43.80	57.83	37.72				-4%	27%	-17%
12	Avista Corp	5	0%	0%	8%	0%	92%	0.034	0.028	0.028	0.018	82%	81%	54%	3.37	2.96	4.49	2.81				-12%	33%	-17%
13	El Paso Electric Co	8	10%	0%	32%	58%	0%	0.284	0.265	0.237	0.146	93%	83%	51%	23.53	19.08	31.39	16.96				-19%	33%	-28%
14	Los Angeles City	13	35%	0%	0%	0%	0%	0.596	0.549	0.450	0.312	92%	76%	52%	30.87	24.74	43.85	24.67				-20%	42%	-20%
15	FPL Group, Inc	124	7%	17%	44%	26%	6%	0.395	0.361	0.300	0.219	91%	76%	55%	30.97	30.93	45.54	32.11				0%	47%	4%
16	Sempra Energy	20	19%	0%	66%	15%	0%	0.461	0.444	0.378	0.263	96%	82%	57%	34.24	31.42	51.56	33.84				-8%	51%	-1%
17	KeySpan Corp	16	0%	77%	23%	0%	0%	0.758	0.661	0.494	0.332	87%	65%	31%	57.36	61.96	89.64	41.08				8%	56%	-28%
18	Energy Corporation	118	13%	5%	18%	63%	0%	0.279	0.285	0.245	0.177	102%	88%	63%	17.58	16.62	28.08	18.49				-5%	60%	5%
19	CLECO Corporation	9	45%	0%	55%	0%	0%	0.735	0.679	0.567	0.422	93%	77%	57%	38.40	33.65	61.85	42.67				-12%	61%	11%
20	Reliant Resources	41	52%	0%	40%	0%	1%	0.768	0.701	0.605	0.414	91%	79%	54%	30.81	25.88	54.33	32.65				-16%	76%	6%
21	Seminole Electric Coop Inc	11	59%	23%	0%	0%	0%	0.786	0.757	0.708	0.583	96%	90%	74%	36.50	32.28	64.49	53.03				-12%	77%	45%
22	Jacksonville Electric Auth (JEA)	11	54%	39%	7%	0%	0%	0.928	0.876	0.875	0.736	94%	94%	79%	43.45	39.21	78.81	68.72				-10%	81%	58%
23	Pepco Holdings, Inc	7	52%	12%	36%	0%	0%	0.824	0.762	0.676	0.545	93%	82%	66%	32.69	30.02	61.05	46.37				-8%	87%	42%
24	Mirant Corp	36	57%	74%	73%	0%	0%	0.916	0.746	0.692	0.810	91%	75%	66%	31.93	29.76	64.08	44.41				-7%	100%	10%
25	Public Service Enterprise Group (PSEG)	52	20%	4%	28%	47%	0%	0.346	0.362	0.317	0.239	105%	92%	69%	17.98	18.01	34.18	23.86				0%	90%	33%
26	Enron Corporation*	6	44%	28%	28%	0%	26%	0.603	0.566	0.467	0.371	94%	77%	62%	22.13	20.51	42.64	31.95				-7%	93%	44%
27	TECO Energy	20	53%	1%	46%	0%	0%	0.702	0.671	0.578	0.444	96%	82%	63%	28.62	26.54	55.16	39.32				-7%	93%	37%
28	Lower Colorado River Authority	12	58%	0%	39%	0%	3%	0.734	0.695	0.602	0.465	95%	82%	63%	29.92	27.10	57.92	42.35				-9%	94%	42%
29	Sierra Pacific Resources	13	65%	0%	34%	0%	2%	0.834	0.800	0.718	0.565	95%	86%	64%	29.54	25.88	57.71	40.96				-12%	95%	39%
30	AES Corp	43	75%	3%	20%	0%	2%	0.865	0.788	0.730	0.546	91%	84%	63%	25.35	21.01	51.59	33.97				-17%	104%	34%
31	Austin Energy	10	41%	0%	24%	34%	0%	0.537	0.548	0.484	0.380	102%	90%	71%	22.84	21.98	46.96	35.62				-4%	100%	56%
32	OGE Energy Corporation	22	72%	0%	28%	0%	0%	0.898	0.818	0.752	0.598	91%	84%	67%	29.03	25.96	62.27	46.73				-11%	114%	61%
33	Dynegy Inc	33	69%	12%	19%	0%	0%	0.874	0.812	0.767	0.613	93%	88%	70%	29.79	27.01	65.26	51.49				-9%	119%	73%
34	Exelon Corporation	151	5%	1%	2%	91%	1%	0.072	0.085	0.070	0.056	119%	98%	78%	2.44	2.67	5.35	3.93				9%	119%	61%
35	Hawaiian Electric Industries, Inc	7	0%	100%	0%	0%	0%	0.804	0.795	0.796	0.749	99%	99%	93%	61.72	59.17	145.10	135.92				28%	135%	120%
36	Texas Genco LLC*	46	68%	0%	17%	14%	0%	0.766	0.735	0.686	0.549	96%	90%	72%	24.86	22.86	58.52	46.07				-8%	135%	85%

29

#	Company	C1	C2	C3	C4	C5	C6	C7	C8	C9	C10	C11	C12	C13	C14	C15	C16	C17	C18	C19
37	NRG	28	80%	5%	15%	0%	0.930	0.838	0.811	0.624	90%	87%	67%	25.76	22.83	60.87	45.52	-11%	136%	76%
38	Goldman Sachs	14	71%	1%	28%	0%	0.836	0.778	0.716	0.573	93%	86%	69%	24.37	22.94	58.01	44.27	-6%	138%	83%
39	Pinnacle West Capital Corporation	26	48%	0%	20%	32%	0.557	0.570	0.511	0.402	102%	92%	72%	17.61	17.26	42.24	31.26	-2%	140%	78%
40	Xcel Energy Inc	81	67%	1%	16%	16%	0.780	0.741	0.693	0.554	95%	89%	71%	20.18	19.03	37.65	37.65	-6%	149%	87%
41	Progress Energy	93	47%	8%	11%	33%	0.573	0.584	0.530	0.425	102%	92%	74%	15.40	16.14	50.21	29.69	5%	150%	93%
42	CMS Energy Corporation	30	62%	1%	17%	18%	0.707	0.690	0.626	0.477	98%	89%	67%	19.14	18.99	38.52	35.84	-1%	153%	87%
43	Duke Energy Corporation	102	44%	0%	16%	38%	0.442	0.481	0.419	0.300	109%	95%	68%	12.74	13.55	48.34	32.36	6%	153%	67%
44	San Antonio Public Service Bd	20	54%	0%	16%	30%	0.652	0.647	0.591	0.477	99%	91%	73%	18.12	18.33	46.08	35.28	1%	154%	95%
45	Northeast Utilities	7	67%	24%	0%	8%	0.923	0.852	0.840	0.689	92%	91%	75%	21.52	21.52	54.43	45.13	1%	155%	112%
46	Alabama Electric Coop Inc	6	77%	0%	23%	0%	0.923	0.838	0.793	0.648	91%	86%	70%	21.41	20.70	56.23	43.14	-3%	163%	101%
47	Constellation Energy Group, Inc	56	31%	2%	10%	2%	0.337	0.381	0.330	0.262	113%	98%	78%	9.76	11.00	25.73	19.47	13%	164%	100%
48	Salt River Proj Ag I & P Dist	25	65%	0%	17%	55%	0.709	0.693	0.643	0.518	98%	91%	73%	17.37	17.40	46.99	35.81	-3%	170%	106%
49	Alcoa, Inc	9	71%	0%	4%	20%	0.670	0.645	0.565	0.452	96%	84%	67%	14.79	14.78	41.97	33.39	0%	184%	126%
50	TXU Corporation	67	65%	0%	6%	25%	0.734	0.716	0.688	0.567	97%	94%	77%	20.18	19.01	50.21	45.99	-2%	184%	136%
51	WPS Resources Corporation	14	77%	1%	2%	0%	0.873	0.776	0.757	0.515	89%	87%	59%	17.69	16.79	51.43	35.99	-5%	191%	103%
52	UniSource Energy Corporation	11	96%	0%	4%	3%	0.109	0.112	0.073	0.061	103%	67%	56%	1.38	1.80	4.03	3.30	30%	192%	139%
53	E.ON (Germany)	47	92%	7%	0%	1%	0.963	0.883	0.914	0.752	92%	95%	78%	19.53	19.03	57.64	47.61	-3%	195%	144%
54	Southern Company Inc	178	70%	0%	10%	3%	0.717	0.702	0.656	0.518	98%	94%	72%	14.60	15.25	43.64	33.04	4%	199%	126%
55	Westar Energy	16	62%	5%	16%	29%	0.717	0.716	0.693	0.566	100%	97%	79%	15.89	16.52	47.98	37.46	0%	202%	136%
56	MidAmerican Energy Holdings Co	24	75%	0%	3%	13%	0.754	0.724	0.724	0.595	96%	92%	79%	17.47	17.11	52.88	43.56	1%	203%	149%
57	Alliant Energy Corp	24	73%	2%	7%	4%	0.789	0.754	0.724	0.595	96%	92%	75%	17.47	16.52	52.46	43.56	-1%	206%	148%
58	PPL Corp	50	52%	0%	3%	21%	0.810	0.766	0.755	0.609	95%	93%	75%	17.17	16.93	52.46	42.53	1%	211%	148%
59	Great River Energy	9	98%	5%	3%	96%	0.534	0.471	0.483	0.383	100%	94%	76%	9.54	11.22	29.64	23.67	18%	212%	174%
60	IDACORP Inc	13	54%	0%	7%	0%	1.116	0.972	1.030	0.871	87%	92%	78%	26.38	24.22	82.40	72.21	-8%	213%	161%
61	Dominion	102	49%	4%	0%	46%	0.554	0.533	0.433	0.361	96%	78%	65%	8.49	9.68	22.14	22.14	14%	214%	143%
62	Associated Electric Coop Inc	17	91%	0%	9%	0%	0.526	0.557	0.507	0.400	100%	96%	76%	10.58	12.25	33.22	25.72	16%	215%	150%
63	Edison International	78	70%	0%	0%	1%	0.914	0.845	0.844	0.689	92%	92%	75%	19.18	19.14	60.40	49.19	0%	219%	160%
64	Santee Cooper (SC Public Service Auth)	24	78%	2%	5%	5%	0.717	0.704	0.668	0.546	98%	93%	76%	14.21	15.25	45.34	36.88	7%	225%	157%
65	ScottishPower	50	87%	1%	7%	1%	0.853	0.803	0.791	0.656	94%	93%	77%	14.38	15.44	46.78	37.01	7%	226%	164%
66	NiSource, Inc	16	93%	2%	5%	7%	0.909	0.830	0.830	0.672	91%	92%	74%	16.15	16.46	52.60	42.69	2%	228%	171%
67	Arkansas Electric Coop Corp	11	89%	4%	0%	6%	1.037	0.923	0.956	0.801	89%	92%	77%	18.51	18.51	60.64	50.22	2%	230%	180%
68	TransAlta	11	90%	0%	2%	7%	0.957	0.869	0.881	0.738	91%	92%	77%	17.80	18.11	58.78	49.89	0%	230%	179%
69	U S Bureau of Reclamation	41	11%	0%	0%	89%	0.929	0.846	0.844	0.750	91%	91%	81%	17.96	17.20	59.33	50.17	-1%	234%	171%
70	ALLETE	9	92%	0%	0%	8%	0.946	0.857	0.847	0.663	91%	90%	70%	16.45	16.96	55.63	44.22	3%	238%	169%
71	Municipal Electric Authority of Georgia	13	41%	0%	6%	53%	0.427	0.471	0.424	0.346	110%	99%	81%	8.57	10.36	29.03	23.06	21%	239%	169%
72	Basin Electric Power Coop	16	100%	0%	0%	0%	0.975	0.975	1.043	0.884	100%	107%	91%	21.80	21.01	74.58	64.74	0%	242%	197%
73	American Electric Power Co Inc	190	83%	0%	4%	12%	0.781	0.759	0.759	0.582	97%	97%	74%	13.82	15.00	47.39	42.24	-4%	243%	163%
74	Oglethorpe Power Corp	22	48%	0%	4%	46%	0.490	0.527	0.479	0.393	108%	98%	80%	9.16	10.98	31.85	25.65	9%	248%	180%
75	PNM Resources Inc	10	70%	0%	28%	2%	0.677	0.684	0.542	0.462	101%	80%	68%	12.54	13.76	44.04	36.07	10%	251%	188%
76	Ameren Corp	75	87%	0%	10%	2%	0.796	0.793	0.625	0.542	101%	80%	68%	14.67	15.83	52.36	41.98	8%	257%	186%
77	Cinergy Corp*	65	95%	0%	4%	1%	0.884	0.879	0.696	0.625	99%	79%	71%	14.45	15.60	51.62	40.17	2%	257%	178%
78	Omaha Public Power District	12	66%	0%	33%	0%	0.648	0.661	0.637	0.527	102%	98%	81%	12.00	13.60	43.41	36.46	13%	262%	204%
79	DTE Energy Company	49	80%	1%	17%	1%	0.778	0.750	0.740	0.583	97%	96%	76%	12.59	14.15	46.36	36.53	12%	268%	190%

	Company															
80	Great Plains Energy Inc	21	76%	0%	1%	23%	0%	0.823	0.785	0.663	95%	96%	81%	15.17	50.89	43.01	10%	269%	212%
81	Wisconsin Energy Corporation	29	70%	0%	0%	27%	2%	0.734	0.720	0.588	98%	96%	80%	12.21	13.82	45.28	38.06	13%	271%	212%
82	FirstEnergy Corporation	79	60%	1%	0%	38%	0%	0.573	0.604	0.456	105%	99%	80%	9.08	11.08	33.77	26.80	22%	272%	195%
83	Tri-State G & T Assn Inc	11	99%	0%	1%	0%	0%	1.036	0.925	0.816	89%	94%	79%	15.99	16.88	60.26	50.34	6%	277%	215%
84	East Kentucky Power Coop Inc	9	97%	0%	2%	0%	1%	0.922	0.855	0.719	93%	95%	78%	13.68	15.14	51.60	41.59	11%	277%	204%
85	SCANA Corporation	24	73%	0%	3%	23%	1%	0.685	0.692	0.538	101%	97%	79%	10.21	12.33	38.64	30.15	21%	278%	195%
86	Nebraska Public Power District	17	63%	0%	0%	36%	0%	0.694	0.690	0.567	99%	97%	82%	11.16	12.92	42.59	36.63	16%	282%	233%
87	Vectren Corporation	7	99%	0%	1%	0%	0%	1.039	0.926	0.815	89%	94%	78%	14.46	15.74	56.21	46.15	9%	289%	219%
88	Hoosier Energy R E C Inc	8	99%	0%	0%	0%	0%	0.984	0.893	0.773	91%	95%	79%	13.82	15.33	54.23	44.23	11%	292%	220%
89	DPL Inc	17	99%	0%	0%	0%	0%	0.889	0.835	0.673	94%	96%	76%	12.77	14.52	50.30	38.92	14%	294%	205%
90	Tennessee Valley Authority	158	61%	0%	0%	30%	9%	0.597	0.609	0.474	100%	93%	71%	8.90	10.16	31.65	23.77	27%	180%	197%
91	Allegheny Energy Inc	43	97%	0%	2%	0%	0%	0.957	0.877	0.777	100%	100%	70%	12.30	14.30	43.38	34.31	10%	305%	215%
92	Buckeye Power Inc	9	100%	0%	0%	0%	0%	0.868	0.826	0.686	95%	97%	79%	12.00	14.06	49.22	39.40	17%	310%	228%
93	Intermountain Power Agency	14	100%	0%	0%	0%	0%	1.009	0.909	0.798	90%	95%	79%	11.72	13.88	50.95	41.11	18%	335%	251%

* Companies were acquired by other companies and/or transferred to a newly entity but in the analysis are still considered separate companies: (1) Cinergy Corporation was acquired by Duke Energy Corporation, (2) Delta Power was largely acquired by Arroyo Energy and the rest transferred to the new entity named Olympus Power, (3) Enron's power generation unit (Prisma Energy International) was acquired by Ashmore Energy International, and (4) Texas Genco was acquired by NRG.

Seven companies produced electricity in 2004 purely from carbon-free energy resources and were therefore not included in the carbon performance analysis:

- Center Point Energy
- Energy Northwest
- North Carolina Municipal Power Agency
- PUD of Chelan County
- PUD of Grant County
- Seattle City Light
- US Corps of Engineers

31

Annex

Paper IV

available at www.sciencedirect.com

ScienceDirect

www.elsevier.com/locate/ecolecon

ANALYSIS

Emerging carbon constraints for corporate risk management

Timo Busch*, Volker H. Hoffmann

Group for Sustainability and Technology, Department for Management, Technology, and Economics, ETH Zurich,
Kreuzplatz 5, 8032 Zurich, Switzerland

ARTICLE INFO

Article history:
Received 15 February 2005
Received in revised form
28 April 2006
Accepted 3 May 2006
Available online 25 September 2006

Keywords:
Climate change
Disposition of fossil fuels
Carbon exposure
Risk management

ABSTRACT

While discussions about global sustainability challenges abound, the financial risks that they incur, albeit important, have received less attention. We suggest that corporate risk assessments should include sustainability-related aspects, especially with relation to the natural environment, and encompass the flux of critical materials within a company's value chain. Such a comprehensive risk assessment takes into account input- as well as output-related factors. With this paper, we focus on the flux of carbon and define carbon constraints that emerge due to the disposition of fossil fuels in the input dimension and due to direct and indirect climate change effects in the output dimension. We review the literature regarding the financial consequences of carbon constraints on the macroeconomic, sector, and company level. We conclude that: a) financial consequences seem to be asymmetrically distributed between and within sectors, b) the individual risk exposure of companies depends on the intensity of and dependency on carbon-based materials and energy, and c) financial markets have only started to incorporate these aspects in their valuations. This paper ends with recommendations on how to incorporate our results in an integrated carbon risk management framework.

© 2006 Elsevier B.V. All rights reserved.

1. Introduction

In the scientific literature, there are a number of studies that examine the relationship between the financial performance of companies and their efforts towards sustainability (e.g., King and Lenox, 2001; Murphy, 2002; Orlitzky et al., 2003; Stanwick and Stanwick, 1998; Margolis and Walsh, 2001). Despite some methodological concerns (Vogel, 2005), the majority of investigations suggests that there is a positive – or at least not negative – correlation. In order to understand whether and how efforts towards sustainability can contribute to a superior financial performance, it is crucial to analyze corresponding business risks. With this paper, we suggest systematically integrating risks that are related to the natural environment[1]

into corporate risk management by focusing on the flux of critical materials within the value chain of a company.

Traditional risk assessments typically focus on tangible uncertainties for which probabilities can easily be quantified and traced back. With respect to the natural environment, risk assessments so far have emphasized adverse impacts and financial consequences for companies when they release materials or emissions (e.g., Joshi et al., 2005). In contrast to this, other more intangible and rather unpredictable factors such as ecological limitations receive less attention as definite scientific proof might be lacking or precise probabilities of occurrence are difficult to estimate. But as political framework conditions, consumption patterns, and global markets change, some of these factors should be considered from a different point of view.

---

* Corresponding author. Tel.: +41 44 632 05 53
  E-mail address: tobusch@ethz.ch (T. Busch).
[1] With the term 'environment,' we refer to the business environment of a company; when we refer to the natural environment, this will be mentioned explicitly.

0921-8009/$ - see front matter © 2006 Elsevier B.V. All rights reserved.
doi:10.1016/j.ecolecon.2006.05.022

New constraints emerge that are likely to affect the financial performance of companies, either directly by influencing their operations or indirectly by influencing the value chain they operate in. It is a central idea of this paper that the way in which companies assess risks that are related to the natural environment should be extended to reflect such emerging constraints in the input and output dimension. A similar development occurred in corporate environmental management: until the early 1990s, there was a clear focus on end-of-pipe technologies in order to reduce emissions. Later, however, with the upcoming discussion of eco-efficiency, the focus switched to comprehensive approaches such as integrated production and pollution prevention. For example, McDonough and Braungart (2002) have suggested analyzing the whole life cycle of a product or production process. In order to reduce toxic product features, they focus on rethinking what products are made of. Analogously, in order to comprehensively address risks that stem from the natural environment, all different kinds of risks along the value chain of a company should be taken into account.

With this paper, we discuss emerging constraints for companies related to the element carbon. As the detailed analysis of all carbon-related constraints would go beyond the scope of this paper, we make three restrictions to our analysis. First, to illustrate carbon constraints in the input dimension, we focus on the availability conditions of crude oil as the main carbon input factor (IEA, 2004). However as all fossil fuels are finite, our scarcity considerations can be applied to other fossil fuels as well. Second, the output-oriented carbon constraints that are discussed in this paper include carbon dioxide (CO_2), while other greenhouse gases (GHG) are not considered. Nevertheless, carbon dioxide is the key anthropogenic GHG and, therefore, an important source of man-made climate change (Baumert et al., 2005; Intergovernmental Panel on Climate Change, 2001). Third, technological change and the option of accelerating the use of alternative fuels and other substitutes are not considered explicitly. This has been discussed by Bretschger (2005) in the context of resource scarcity and by Gerlagh and Lise (2005) in the context of carbon taxes. It is, therefore, an assumption – albeit also a noteworthy limitation – that the underlying investigations and statistics cited in this paper have sufficiently embraced these factors.

To build on our idea of integrating the flux of critical materials in the risk management of companies, we proceed in four steps. First, this paper defines carbon constraints from a corporate risk management perspective (Section 2). Next, it scrutinizes carbon constraints that are related to resource inputs, namely by elaborating on disposition aspects of fossil fuels (Section 3). Subsequently, the paper analyzes carbon constraints that are related to the output dimension, namely by delineating the discussion about climate change (Section 4). Sections 3 and 4 each focus on reviewing the literature on financial implications of input and output related constraints, respectively. Finally, the findings are discussed and recommendations are derived for an integrated carbon risk management framework (Section 5).

2. Definition of carbon constraints

Clark et al. (2001) describe a framework to characterize and classify information about global environmental issues and their management. One important aspect of this framework is the flux of materials and its effects on the natural environment. When examining the flux of carbon and its implications that are relevant for business, we differentiate between effects within and outside of a company's own operations. A company is considered as a single entity that requires carbon-based inputs (e.g., oil) and releases carbon containing outputs (e.g., CO_2). These inputs and outputs are subject to a variety of influences (e.g., new regulations) that can affect the financial performance of a company. However, these influences can financially affect a company even if they occur not in the company itself, but within the value chain of the company. In order to identify and incorporate all life-cycle-wide effects, we focus on the flux of critical materials within the entire value chain.

The general business environment of a company can be characterized by uncertainty, complexity, and munificence (Aragon-Correa and Sharma, 2003). These attributes also apply to the flux of critical materials within the value chain of a company or – specifically in the focus of this paper – to the implications of utilizing carbon-based materials and energy. In the context of risk management, we focus on the aspect of uncertainty. In order to derive and assess financial risks for a company, the probability and distribution of uncertain parameters have to be considered (Morgan and Henrion, 1995). This especially holds for parameters that have the potential to restrict – or constrain – the actions of corporate management. We define constraints as all sorts of conditions that limit companies in their way of conducting business and in their efforts towards profitability. If new constraints emerge, management strategies are required in order to adapt established processes or to mitigate the constraints' sources. Focusing on carbon issues, we call these constraints *carbon constraints*. In the following section, we distinguish carbon constraints on the input and output side of a company as exemplified in Fig. 1.

Carbon constraints on the input side of the value chain are related to the disposition of fossil fuels. These are determined and influenced twofold, by natural scarcity as well as by sociopolitical factors. On the one hand, as fossil fuels are naturally limited, scarcity and the resulting market reactions have to be evaluated. Resource scarcity is determined by several factors such as endowment of natural stocks, availability of resources and reserves, and technical developments that are able to postpone the final depletion. The resulting adjustment processes on markets influence prices; therefore, possible price reactions of the market have to be scrutinized. On the other hand, factors such as government-related measures (e.g., taxes), international political developments (especially in oil-producing countries), and change in consumer preferences have to be contemplated to additionally influence price developments.

Carbon constraints on the output side of the value chain can be divided into direct and indirect climate change effects. The former describe direct physical impacts of climate change on a company's assets and processes such as damage to production facilities or availability of raw materials (e.g., water). Impacts on human health also have to be considered in this context. The latter encompass impacts due to global warming mitigation efforts, such as the European CO_2-

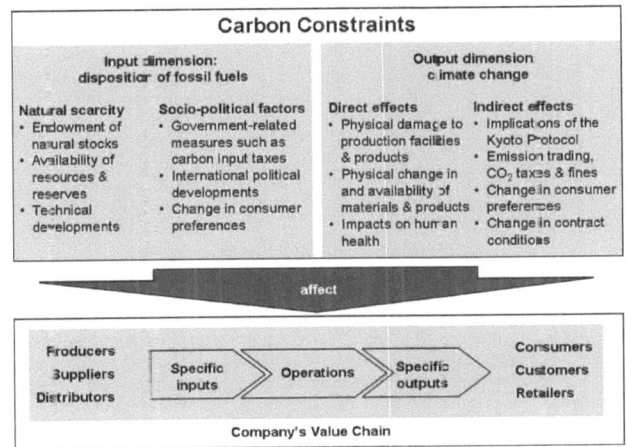

Fig. 1 – Definition of carbon constraints for companies.

Emission Trading Scheme (EU ETS) or implications of the Kyoto Protocol. Furthermore, aspects such as a change in contract conditions (e.g., insurances) and a change in consumer preferences have to be contemplated.

Both input and output constraints affect the fabrication of materials (e.g., carbon-based plastics) as well as energy prices in general. This consideration stresses the importance for companies to take into account the entire value chain – not only internal activities have to be evaluated but suppliers and customers can also be subject to higher risks and, therefore, affect the financial bottom line. In order to derive an industry's or a company's exposure to carbon constraints, the intensity of and dependency on carbon inputs and outputs as well as corresponding market developments have to be scrutinized.

3. Carbon constraints from an input perspective

This chapter first discusses the disposition of fossil fuels in the light of uncertainty; next, the current discussion focusing on how resulting constraints are likely to affect financial performance is presented, followed by concluding remarks.

3.1. Disposition of fossil fuels – an emerging issue

Uncertainties due to the disposition of fossil fuels encompass all factors that could possibly disrupt the supply demanded. We distinguish between two different ways in which the disposition of fossil fuels can be considered as a constraint for companies: natural scarcity and socio-political factors.

In order to illustrate carbon constraints in the input dimension, we focus on the disposition of oil and its natural restrictions.[2] The risk of natural oil scarcity is defined as a risk

[2] For an overview of global resources and reserves of all fossil fuels, see Federal Institute for Geosciences and Natural Resources (2005).

derived from limited resources due to exploitation and restricted access (World Resource Institute, 2002). Estimations for the cumulative production worldwide (all technically possible explorations over time) range between 1850 and 2200 billion barrels (Association for the Study of Peak Oil and Gas, 2005; Meacher, 2004). Campbell (1997) refers to reserves as the amounts of petroleum that are estimated to be recoverable from a known petroleum accumulation as of a stated reference date based on economic and technological assumptions. The estimations for reserves range between 905 and 1189 billion barrels (Association for the Study of Peak Oil and Gas, 2005; British Petroleum, 2005). Despite this rather distinct indication for a finite resource, an examination of the demand side illustrates a steady increase: between 1984 and 2004 oil consumption worldwide increased by 56% to a level of 83.4 million bbl/day in 2005 and a further increase of 38% within the next 25 years can be expected (based on British Petroleum, 2005; IEA, 2005a,b). Taking into account the estimated reserves and assuming a linear development of the increase, in 2030 between 58% and 76% of 2005 reserves would be depleted. However, it is not of practical use in the context of this paper to analyze when the ultimate barrel will be exhausted. In light of the corporate constraint perspective, what is more important is when markets are going to acknowledge the upcoming scarcity.

The significance of upcoming resource scarcity can be analyzed with physical and economic indicators. Regarding physical measures, Hubbert (1974) explains oil discovery and production as a bell-shaped curve; it illustrates that the decline in oil production flattens over time. The peak of the Hubbert-curve is defined as depletion mid-point. In theory, the production peak-point and depletion mid-point should coincide. But production can be held up (or even increased) artificially for a few years through technical interference or quota regulation. As a result, there is no unanimous opinion in public debates: the IEA (2005b) states that "global oil reserves exceed the cumulative projected production between now and

2030, but reserves need to be proved up in order to avoid a peak in production before the end of the projection period." The Association for the Study of Peak Oil and Gas (2005) states that the peak production of regular oil was already reached in 2004. Campbell (1997) estimated the depletion mid-point for 2000, Odell (2004) forecasts 2011. Due to the ambiguous picture based on physical indicators, we conclude that there is no inducement for markets to reflect on oil scarcity as an upcoming issue from a short-term perspective.

Beyond physical measures, economic indicators can be utilized in order to deliver more detailed information. Solow (1974) summarizes the theoretical discussion about optimal extraction of resources and refers to Hotelling (1931): different resources are used in the order of their specific extraction costs which can vary depending on geographical accessibility, related transportation costs or geological differences. The "marginal cost of resource extraction, together with the opportunity cost of a unit of resource in the ground, helps in setting the price [...]. As extraction proceeds, the opportunity cost of a unit of the resource rises" (Ruth, 2006). As a result, under the assumption of constant or increasing demand, the price should be negatively correlated to the size of the stock. But Barnett and Morse (1963) as well as Scott and Pearse (1992) show that resource prices have not provided any significant evidence for scarcity. A similar indication was reached in the Simon–Ehrlich debate in which the price vs. scarcity topic was broadly discussed (Bureau of Transport and Regional Services, 2005; Tierney, 1990). Despite declining physical resources – a basket of five metals was considered – the inflation-adjusted price level in 1990 was below the starting point in 1980. As a rather new approach to analyzing price developments, Reynolds (1999) suggests that exploration costs constitute an adequate measure for scarcity. He assumes that the explorer never entirely knows the size of the resource base. Exploration costs decrease when the explorer gains new information about new reserves as discovery proceeds. As an effect, the true scarcity will not be revealed until close to exhaustion. This could result in a sudden and sharp increase in resource prices, even after decades of declining prices (see also Bakhtiari, 2004; Tierney, 1990). Supporters of new institutional economics suggest path-dependencies as one possible explanation for inefficiencies on resource markets. One reason for path-dependencies could be sunk costs associated with the difference in the value for assets, ex ante and ex post their purchase (Antonelli, 1997). This circumstance can lead to less commitment in the development of large technological innovations (Walker, 2000). This could delay the substitution of carbon-based inputs by renewable resources. Following these points of view regarding economic indicators, we conclude that the actors' behaviors on markets appear to be biased, at least from a short-term perspective. However, we argue that once natural scarcity has been acknowledged as an emerging issue from a long-term perspective, this will prompt markets to incorporate higher risk premiums and will result in an adjustment of prices. This is likely to also affect price volatilities (Hirsch et al., 2005; Lovins et al., 2005; Sadorsky, 1999).

In addition to these natural scarcity considerations, socio-political factors affect the disposition of carbon inputs and, thus, influence prices and their volatility. On the one hand, with respect to legislation, carbon taxes have to be considered which are applied directly on the use of carbon emitting inputs. In this context, Wolfson and Koopmans (1996) suggest a timely introduction of regulatory taxation in order to reduce the demand for fossil fuels and accelerate the introduction of more sustainable technologies. On the other hand, fossil fuels are subsidized by governments, in the case of the EU by about EUR 24 billion in 2004 (Allianz Group, WWF, 2005). Lowering the amount of subsidies is similar to introducing a new tax, therefore this effect also has to be considered when examining input-related carbon constraints.

3.2. Financial implications of oil scarcity

We now review studies analyzing the financial implications of oil scarcity on the macroeconomic, sector, and company level. On the *macroeconomic level*, oil plays a crucial role in economic development (IEA, 2004; The Economist, 2005a) and in terms of energy supply as crude oil currently has a share of 35% of the world's primary energy demand (IEA, 2005b). This can be illustrated by a recent report by Goldman Sachs (2006) who surveyed economists worldwide to gauge perceptions of economic growth and assess the threats that the global economy faces. When asked to assess the probability of ten specific risks to the global economy, economists put world oil supply at the top of the list. However, the financial implications of oil scarcity are discussed controversially and are difficult to quantify as a number of uncertainties are eminent. First, consumption can still increase, thereby accelerating the natural depletion effect; the current rather low level of certainty could be maintained from a short-term perspective. However from a long-term perspective, uncertainty is significant because "if economic growth remains technologically dependent on growth in conventional oil supply until global conventional oil production peaks, then the development following the peak is unpredictable" (The Danish Board of Technology, 2003). Second, technological development is often seen as a means of extending the availability of oil into the future. An important question relates to how technological development will impact supply and demand (Bureau of Transport and Regional Services, 2005). While technology development seems hard to predict and large uncertainties remain, it is likely that financial implications on the macroeconomic level will be significant. For example, the Union of Concerned Scientists (2002) found that the costs of oil dependency to the US economy were as high as USD 7 trillion over the past three decades; by fostering the usage of fuel-efficient vehicles, buildings, and industry, oil savings could total 5.9 million bbl/day by 2020. This would equal annual consumer savings of USD 150 billion.

Studies regarding the *sector level* provide deeper insights: the IEA (2004) points out that the manufacturing and services industries as well as households decreased their oil demand (total effect of IEA-11 countries; measured in average annual percent change; 1973–1998), whereas the travel and freight industries increased their oil demand. The latter two sectors have very high exposure, as crude oil provides almost 90% of all transportation fuel (Roberts, 2004). Focusing on socio-political factors and using the US pulp and paper industry as an example, Ruth et al. (2000) illustrate the potential impacts of policies that increase the cost of carbon-containing inputs. The authors analyze a policy-induced price increase and show that energy expenditures are likely to decrease in the case of an additional investment incentive for new technologies. In a later report Ruth et al. (2004) conducted three case studies, namely for the pulp and paper, iron and steel, and ethylene industries in the US. Their findings illustrate how carbon input and investment policies can affect the various industries differently. Cost increases in carbon-emitting inputs have a relatively high impact on the steel and paper industry; the net carbon intensity (the use of carbon-emitting inputs per unit of output) is estimated to be reduced significantly. Compared to that, effects on the petrochemical industry are smaller but depend on the type of policy that is applied. In this context, Vellema et al. (2003) state that a general concern of the chemical industry is "the intense use of finite resources such as oil." Based on a general description and various case studies, the authors illustrate how enhanced technologies will facilitate the incorporation of sustainability in fields such as the production of monomers and polymers or efficient waste management.

On the *company level*, it can be illustrated that unstable oil prices already present a severe risk (Fishhaut, 2003). Emerging constraints due to oil scarcity can be described in two dimensions. In the short run, price fluctuations affect the distribution of oil and can, thus, disturb or interrupt individual production processes. However, ensuing supply delays or reduced production capacities are just temporary issues. In the long run, the effects of upcoming oil scarcity constitute an uncertain business environment. This impedes accurate and reliable investment planning for both managers and investors. Even though upcoming scarcity might not be displayed in actual price developments, an indication of increasing volatility of oil prices can be seen in the increased trading volume of crude oil future contracts (Dobbs et al., 2006) which is also confirmed by other sources (e.g., Balls and Giles, 2005; The Economist, 2005b). In addition, capital markets seem to be sensitive regarding carbon regulations. This can be illustrated by the example of Xstrata, a FTSE 100 listed company that exports coal to Japan. In 2002, the Japanese government announced that it was considering a coal levy. The shares of Xstrata fell approximately 10% in just a few days, due to the anticipated negative impact of such a regulation (UNEP Finance Initiative, 2003; Kierman, 2002).

3.3. Concluding results: carbon inputs as constraint

On the macroeconomic level, oil markets have priced mainly socio-political factors so far: significant price movements have been determined by artificial scarcity, due to political conflicts, economic interventions or supply restrictions. Regarding price developments there seem to be discrepancies between the Hotelling principle and the actual scarcity patterns. We conclude that fossil fuel markets are rather short-term focused and current prices can be subject to ongoing path-dependencies. This can result in a sudden price increase, unpredicted shortages, and high market uncertainty. From a rather long-term perspective, scarcity will emerge sooner or later after oil production has peaked (Campbell, 1997; Meacher, 2004; Roberts, 2004). On the sector level, the economic and financial implications of such an abrupt allocation problem can be enormous and are likely to affect different sectors to a different extent. The crucial question is when this will become tangible for which industries. In the short run, the price elasticity of oil can be regarded as quite low because the substitution of assets using oil as an input factor is time-consuming and requires financial resources. As a result, industries will rely on carbon inputs for a certain time. In the long run, the development of new, oil-independent technologies and the use of alternative energy sources are necessary. On the company level, the understanding of increasing oil prices as a risk factor seems to be rather unsystematic; natural scarcity seems not to be regarded as a short-term issue. But as typical depreciation periods of large scale investment projects usually consider a time frame of 20–30 years, the incorporation of this issue now seems advisable (Lovins et al., 2005). Referring to the example of the automotive industry, Merrill Lynch (2005) similarly emphasize that management strategies need to be adjusted now. Notably, the increasingly volatile oil prices require adequate risk assessment and quantification (Cabedo and Moya, 2003). Existing management systems and risk assessments need to be extended, especially to reflect risks accordingly in companies' long-term budget planning (Innovest Strategic Value Advisors, 2002b).

4. Carbon constraints from an output perspective

In the contemporary literature, innumerable papers and reports discuss climate change and related management and policy strategies. This chapter briefly reviews the output perspective, presents the current state of the discussion focusing on financial effects of climate change and summarizes the findings.

4.1. Climate change – an emerging issue

According to the Third Assessment Report of the Intergovernmental Panel on Climate Change (2001), human activities have led to an increase in the earth's temperature in the 20th century; the accumulation of carbon dioxide and other GHG in the atmosphere is one of the main drivers of climate change. Not only does the actual increase in temperature provide evidence for climate change, but statistics on natural catastrophes, aridity, and flooding are also often cited as an indication of the relevance of this issue (e.g., The Center for Health and the Global Environment, 2005; Munich Re, 2005; Swiss Re, 2005). However, there are still uncertainties and information gaps when assessing climate change and its impacts (European Environment Agency, 2004). Regarding emerging constraints for companies, we differentiate between direct and indirect effects.

The direct effects of climate change on companies can be described as (derived from Mansley and Dlugolecki, 2001; Garz and Volk, 2003 Innovest Strategic Value Advisors, 2002b): prolonged physical impacts on entire industries or individual production facilities (e.g., ski-resorts cannot be utilized as such any more due to a lack of snow; the sporadic destruction of production facilities and infrastructure through extreme weather events (e.g., storms destroying energy transportation systems); the limited availability of raw materials (e.g., the agricultural sector in Mediterranean countries might suffer from an insufficient water supply); and increased dangers for human health (e.g., the potential spreading of tropical diseases). The visibility of these direct effects has increased steadily; e.g., the global economic losses due to natural catastrophes have increased seven-fold in the last 40 years (Munich Re, 2005).

The direct effects are linked with many indirect effects which are caused by human beings as a result of climate change mitigation and adaptation strategies. These indirect effects of climate change can be described as (derived from Mansley and Dlugolecki, 2001; Garz and Volk, 2003; Innovest Strategic Value Advisors, 2002b): consequences of climate change policies (e.g., mitigation policies such as EU ETS); business risks due to policy changes (e.g., a combustion facility turns out to be unprofitable due to the enforcement of high emission taxes; a change in consumer preferences (e.g., consumers switch to products with a lower effect on climate change); and short-term adjustments of contract conditions (e.g., insurances request higher risk premiums due to high climate change exposure). Similar to the direct effects, the indirect effects have also become increasingly tangible. Beyond the EU ETS, several state governments in the US have also undertaken concrete steps to regulate pollutants that cause climate change (Wellington et al., 2004). Furthermore, it is likely that more taxes related to carbon will emerge in the near future. For example, the European Commission has proposed legislation that would require Member States to shift from car registration taxes to CO_2-based car taxation (Commission of European Communities, 2005).

4.2. Financial implications of climate change

We now review studies analyzing the financial implications of climate change on the macroeconomic, sector, and company level. On the *macroeconomic level*, climate change is a relevant business topic (Stern Review, 2006; World Economic Forum, 2000). One aspect of this is an increase in economic losses due to the direct effects of natural disasters which are doubling every ten years and have reached almost USD 1 trillion over the past 15 years (Innovest Strategic Value Advisors, 2002a). Assuming the current trend continues, the annual losses will come close to USD 150 billion in the next decade (UNEP Finance Initiative, 2003). Another more general stand is taken by Tol (2002a) who estimates the impact of a 1 °C warming on GDP worldwide. By using globally averaged values (i.e., using globally averaged prices to value non-market goods and services), he found that the world's GDP would decrease by 3%. Garz and Volk (2003) calculate the worldwide Market-Value-at-Risk (MVaR) of climate change by using scenarios and a dynamic climate model developed by Nordhaus (1994). Based on these calculations, the MVaR is between USD 210 and 915 billion. If the impact on different regions in the world is analyzed, the conclusions seem ambiguous. Tol (2002b) examined the impact of climate change on GDP in a time frame between 2000 and 2200. He found, for example, that the overall impact on the GDP of the OECD is positive. In contrast to this, Innovest Strategic Value Advisors (2003a) focus on GHG-intensive sectors and emission reduction policies; they found that investors will face greater risks from climate change by investing in the EU, Canada, Japan, and parts of the US than in more carbon-diversified portfolios.

On the *sector level*, the average annual CO_2 emissions increased between 1990 and 1998 in all sectors of IEA-11 countries, categorized into manufacturing, households, travel, services, and freight (IEA, 2004). Given their different contributions to climate change (see e.g., Baumert et al., 2005), the various sectors have a different exposure to the financial implications of climate change. For example, Garz and Volk (2003) use a cross-sectional regression model based on data taken from 49 industries to examine the correlation between climate change and company returns since the Rio Conference. The results show that, on the one hand, climate change is not yet priced in as a risk factor, but, on the other hand, companies with a high exposure to climate change effects have to face a potential valuation discount. Kolk and Pinske (2004) found that the most notable industries perceiving government regulations as a risk factor are oil and gas, mining, metals, and utilities; beyond regulation aspects, changing weather patterns will have the most negative impacts on the insurance, food, and beverage industry. For high impact sectors such as energy and electric utilities, Innovest Strategic Value Advisors (2002b) point out that the threat to shareholder value can represent as much as 15% of the total market capitalization. The automotive, chemicals, coal, electric power, manufacturing, oil and gas, refining, steel, and consumer goods industries will most likely be affected negatively by climate change, though the exposure of individual industry sectors varies considerably. Focusing on weather-sensitive sectors, such as agriculture, the effects of climate change are likely to be mixed, i.e., there would be positive as well as negative impacts (Association of British Insurers, 2005b). Hoffmann and Trautmann (in press) show that the risk perception of specific regulatory actions such as the EU ETS and its impact on environmental strategies varies across industries. Also Reinhaud (2005) conducted a study on the costs associated with the EU ETS on several industries (excluding electricity generation). She differentiates between direct costs as compliance costs, i.e., internal abatement costs plus allowance costs, and indirect costs which are attributed to increasing energy prices. Assuming (a) an average allowance price of EUR 10 per ton CO_2 and (b) a 10% allowance need, this would entail a cost increase of 3.7% in the aluminum, 3.4% in the cement, 1.6% in the newsprint, and 1.3% in the steel industries (basic oxygen furnace).

Several other studies have reinforced these findings, although some of them only focus on individual sectors and do not undertake a cross-sector comparison. For instance, electric utilities constitute one of the most environmentally sensitive sectors in the US economy, also in terms of carbon-output-oriented effects (Repetto and Henderson, 2003); Consequently, climate policy will have important financial consequences for power generating companies; costs are likely to exceed 10% of 2002 earnings (Innovest Strategic Value Advisors and WWF, 2003). Reinhaud (2003) illustrates that the introduction of an emission trading scheme influences investment decisions in the power generation sector. Powell et al. (2003) show that the introduction of new emission charges would have strong negative impacts on the aviation industry. Anastas and Kirchhoff (2002) discuss challenges for the chemical industry becoming more environmentally benign and see carbon constraints as one important issue.

The insurance and reinsurance sector has an inherent exposure to the direct effects of climate change. Mills (2005) shows that from 1980 through 2004 the global economic costs of weather-related natural disasters totaled USD 1.4 trillion, of which USD 340 billion were insured. The Association of British

Insurers (2005a) found similar results for household and property accounts in the UK. Globally, the annual losses from the three major storm types – US hurricanes, Japanese typhoons, and EU windstorms – are estimated to increase to USD 27 billion by the 2080s (Association of British Insurers, 2005b). Mills (2005) states that virtually all segments of the insurance industry are vulnerable to the likely impact of climate change. He elaborates on specific technical risks, such as the increasing frequency of loss events or damage functions that increase exponentially. In this context, the Allianz Group and WWF (2005) state that there could be under-pricing of weather-related risks by as much as 30%. In contrast to this, Tol (1998) points out that the impact of climate change on the profitability of the commercial insurance sector does not have to be severe; commercial insurance companies are capable of shifting the risks to the insurants. However, this requires the insurance sector to have proper and timely information about the consequences of climate change. This seems to be a crucial fact as the misestimation of new risks in past years has led to several bankruptcies in the insurance and reinsurance business (Chichilnisky and Heal, 1996).

On the *company level*, the individual exposure depends on a wide range of different factors such as emission-intensity, energy source mix, geographic location of production facilities, marginal abatements costs, and technology trajectory (Innovest Strategic Value Advisors, 2003a). Cogan (2003) examines the climate change business strategies and governance practices of 20 of the world's biggest corporate GHG emitters. The results indicate that all companies are taking some governance action to respond to climate change, while only a few also treat the issue as a financial issue. The level of their engagement differs greatly: two European oil companies examined have pursued all identified governance action points; on the other hand, three US-based oil companies examined have only pursued a minimum governance effort. These differences among major oil companies are investigated by Kolk and Levy (2001) who observe substantial changes: companies are moving towards supporting the Kyoto Protocol and are taking measures to address climate change. The prominent exception is ExxonMobil, confirmed by Mansley (2002, 2003) who investigates the action ExxonMobil has undertaken concerning the impacts of climate risk. Mansley points out that ExxonMobil is not strengthening its strategic position on this issue to the same extent as its competitors. Furthermore, the company is exposing itself to reputation risks, litigation risks, and risks from sudden policy changes. A more extensive analysis is provided by Innovest Strategic Value Advisors (2003b, 2004, 2005a) analyzing the Carbon Disclosure Project in which all FT 500 Global Index companies are requested to disclose investment-relevant information relating to the risks and opportunities presented by climate change. The 2005 report states that more than 90% of the 354 FT 500 companies that responded consider climate change to pose commercial risks and/or opportunities to their business; 86% reported allocating management responsibility to climate change issues. But, on the other hand, only 51% of the respondents have implemented concrete emission reduction programs and only 45% have established emission reduction targets.

The importance of climate change as a topic for investors, notably institutional investors with a long-term perspective, is emphasized by Innovest Strategic Value Advisors (2002b) among others (Cogan, 2004; Garz et al., 2004; Henderson Global Investors, 2002; Mansley and Dlugolecki, 2001; Wellington and Sauer, 2005; Wellington et al., 2004). Investors face economic risks from damage due to the direct effects of climate change and exposure to indirect effects such as climate change abatement costs. Five main risks and potential impacts on shareholder value are identified: balance sheet risk, market and reputational risk, capital cost risk, operating risk, and business sustainability risk. Scenarios by the World Resource Institute (2002) show that climate policies could create different financial impacts for companies in the oil and gas industry, ranging from a 5% loss in shareholder value to a slight gain. Garz and Volk (2003) illustrate that a high management quality (in terms of climate change) decreases the Beta-risk[3] and increases the chances of growth. This result is confirmed by Innovest Strategic Value Advisors (2005b) who set up a climate leaders' portfolio comprising the 21 best-in-class responses to the questionnaire of the Carbon Disclosure Project. The portfolio outperformed its benchmark of same-sector FT 500 peers from 2002 to 2005 by an annual rate of return of 2.3%. As a result, it seems likely that equity analysts incorporate climate change exposure into companies' credit ratings. For example, in 2002 Lehman Brother downgraded the rating on the oil and gas company Canadian Natural because of uncertainty surrounding the Kyoto Protocol (Lehman Brothers Equity Research, 2002). Standard and Poor also warned of the effects of the EU ETS on the cost structure and profitability of European utility corporations (Lund, 2003).

4.3. Concluding results: carbon outputs as constraint

On the macroeconomic level, the studies reviewed highlight the consensus that the direct effects of climate change and global economic losses are interlinked. On the sector level, many studies reveal an asymmetric distribution of climate change effects: the risks and related abatement costs differ between sectors. The electric utility, oil and gas, aluminum, automotive as well as cement industries seem to have high exposure. But in spite of these indications, climate change does not yet seem to be an explicitly priced risk factor on markets. Only for sectors that are affected by indirect effects such as the EU ETS, is climate change considered as an additional cost factor. One exception is the insurance sector: it seems to be broadly accepted that direct climate change effects contribute to higher risk exposure of the entire sector. On the company level, there are indications that financial markets are starting to incorporate climate change in order to determine risk premium rates. However, better performance could also be ascribed to the fact that climate change risk management can be seen as a proxy for overall management capacity (Innovest Strategic Value Advisors, 2005b). We conclude that due to inherent complexity, climate change risks have not been integrated explicitly in the frameworks of risk allocation (Chichilnisky and Heal, 1993). Furthermore, Kunreuther (1978) suggests that companies tend to ignore events with a low probability of occurrence. However, these risks are gaining more and more systematic features and, therefore, there is a growing necessity to consider climate change effects,

---

[3] The Beta-risk is the central risk measurement of the Capital Asset Pricing Model (see e.g., Brealey and Myers, 2000).

for both managers and financial analysts. Individual approaches towards climate change risk management are crucial: companies who manage this issue adequately are increasingly considered to be superiorly managed companies in terms of overall corporate governance.

5. Discussion

In order to adjust established environmental risk assessments to emerging sustainability challenges of the 21st century, we suggest focusing on the flux of critical materials within the value chain of companies. Focusing on the element carbon, this paper discusses that input as well as output-related constraints are going to influence the business environment. There is scientific proof that fossil fuels are on the way to being depleted and climate change is happening. The question is then in how far these developments increase the risk exposure of companies.

In the input dimension, the long-term perspective is that fossil fuels have to be substituted. From a short-term perspective, it is uncertain how and when this will affect a company's profitability. But it can be presumed that once markets have acknowledged the natural scarcity of fossil fuels as an upcoming issue, this will constitute a risk factor for companies. Markets will adjust prices through adequate risk premiums and a further increase in price volatility is very likely. As we illustrated our arguments with the example of crude oil, oil's multiple functions in today's economy have to be emphasized. They range from serving as the source for energy production to making up the basic input factor for numerous production processes; this highlights the wide range of economic sectors that will be affected by an emerging input carbon constraint.

In the output dimension, the long-term implications are subject to uncertainty. On the one hand, there is increasing scientific evidence for climate change and its effects on the economy. But on the other hand, climate-change-related mitigation and adaptation efforts are hard to predict. From a short-term perspective, climate change has started to affect economic development and companies' performance in various ways. Many studies reveal the financial implications of climate change on a macroeconomic level; however, markets will have to foster the exact incorporation of corresponding risks on a microeconomic level.

As a conclusion, we point out three steps companies could take to address emerging carbon constraints within an integrated carbon risk management framework: (1) recognizing the relevance of the topic, (2) determining the company's exposure, and (3) deriving corporate management strategies.

(1) Across all dimensions of our analysis − on the macroeconomic, sector, and company level − it seems plausible that the extent to which different entities will be exposed to carbon constraints varies significantly. All kinds of constraints discussed here are known to a certain extent; however, each of them is subject to a degree of uncertainty as it is dependent on markets pricing the corresponding risks. This highlights the importance of an appropriate analysis of carbon constraints for corporate strategic management. Companies have to recognize that expenditures and revenues of investment projects cannot be calculated exactly as they are subject to uncertainty. This calls for an adequate carbon risk management.

(2) As the next step, companies have to determine carbon constraints for the business sector and asses the company's specific exposure. In general, carbon constraints are asymmetrically distributed. This applies to both the exposure of individual companies within a sector as well as to entire industries; therefore, companies and their value chains have different exposure to the resulting risks (Baumert et al., 2005; Garz and Volk, 2003; Innovest Strategic Value Advisors, 2005a, 2003a; Kolk and Pinske, 2005; IEA, 2004; Repetto and Austin, 2000; Sadorsky, 1999). Similar to the general risk profile, a company's carbon constraint profile is mainly determined by: (a) the company's asset mix, (b) the dependency on and intensity of carbon-based input factors and energy production, (c) the possibility for substitution and technological alternatives, (d) the technological trajectory and industry-specific innovation patterns, (e) the company's position in the value chain, and (f) the location of its operational activities and sales. The detailed assessment of carbon constraints requires the adequate incorporation of various systematic and unsystematic factors, especially the anticipation of potential market reactions. As one main driving force, financial markets and their perception of carbon-related uncertainties and risks have to be analyzed (Bretschger, 2005).

(3) Based on the results of steps 1 and 2, companies are able to derive adequate corporate management strategies in order to respond to the exposure of products, processes, and assets to carbon constraints. Faff and Brailsford (1999), for example, already examined how to add an oil price factor to the Capital Asset Pricing Model in order to value investment strategies. Similarly, a substitution factor could be integrated into established assessment methods showing which assets are likely to be affected by specific input and output risks. Companies are then able to indicate products, processes, and assets which can be substituted or technically altered or which have to rely on carbon-based inputs. The results can help to compare the different carbon exposures with potential alternatives and evaluate different strategies and technologies for both existing as well as future production processes. Notably, proactive management of these issues is necessary especially as companies might have to face carbon constraints not only as a strategic factor for success and competitiveness but for legal reasons as well. For example, the Commission of the German Corporate Governance Code (2005) requests corporate boards of publicly listed companies to disclose any information that is relevant to or could have an impact on the financial situation of the corporation. Following the arguments of this paper, carbon constraints will affect the bottom line and, thus, should be a topic for discussion at future annual general meetings.

For future research, several directions can be identified how scholars can increase the understanding of carbon constraints. In order to incorporate the findings of this paper, an integrated carbon risk management framework has to be developed, and empirically tested, that supports companies in their strategic decision making. Another research direction could focus on the change in technology and the option of accelerating the utilization of alternative fuels. Case studies and scenario analyses, in particular, could help companies to recognize carbon constraints as a business topic and derive their individual exposure. Finally, it was emphasized that markets, and these are most notably financial markets, play a crucial role in determining the relevance of carbon constraints. Scholars should concentrate on the question of how financial markets perceive carbon constraints as a factor that is going to influence the business environment. This would help to reduce the uncertainty in corporate investment planning and, most importantly, might be key to a smooth, i.e., non-disruptive adaptation to carbon constraints – for individual companies as well as for the global economy.

REFERENCES

Allianz Group, WWF, 2005. Climate Change and the Financial Sector: an Agenda for Action, Munich, Gland.

Anastas, P., Kirchhoff, M., 2002 Origins, current status, and future challenges of green chemistry. Accounts of Chemical Research 35 (9), 686–694.

Antonelli, C., 1997. The economies of path-dependence in industrial organization. International Journal of Industrial Organization 15, 643–675.

Aragon-Correa, J.A., Sharma, S., 2003. A contingent resource-based view of proactive corporate environmental strategy. Academy of Management Review 29 (1), 71–88.

Association for the Study of Peak Oil and Gas 2005. Newsletter No. 59, November 2005. Available at: http://www.peakoil.net, 20.12.2005.

Association of British Insurers, 2005a. A Changing Climate for Insurance, a Summery Report for Chief Executives and Policymakers. London.

Association of British Insurers, 2005b. Financial Risks of Climate Change. London.

Bakhtiari, S., 2004. World oil production capacity model suggests output peak by 2006–07. Oil & Gas Journal (26.04.2004).

Balls, A., Giles, C., 2005. Oil Prices and Trade Imbalances Spark Concern at IMF Summit in Washington. Financial Times, London. 26.09.2005.

Barnett, H., Morse, C., 1963. Scarcity and Growth: the Economics of Natural Resource Availability. Johns Hopkins University Press for Resources for the Future Baltimore.

Baumert, K., Herzog, T., Pershing, J., 2005. Climate Data: a Sectoral Perspective. World Resource Institute, Pew Center on Global Climate Change, Arlington.

Brealey, R., Myers, S., 2000. Principles of Corporate Finance, 6th ed. Irwin McGraw-Hill.

Bretschger, L., 2005. Economics of technological change and the natural environment: how effective are innovations as a remedy for resource scarcity? Ecological Economics 54, 148–163.

British Petroleum, 2005. BP Statistical Review of World Energy 2005. London.

Bureau of Transport and Regional Services, 2005. Is the World Running Out of Oil? A Review of the Debate. Australian Government. BTRE Working Paper, vol. 61 Canberra.

Cabedo, D., Moya, I., 2003. Estimating oil price "Value at Risk" using the historical simulation approach. Energy Economics 25 (3), 239–253.

Campbell, C., 1997. The Coming Oil Crisis. Essex.

Chichilnisky, G., Heal, G., 1993. Global environmental risks. Journal of Economic Perspectives 7 (4), 65–86.

Chichilnisky, G., Heal, G., 1996. Catastrophe Futures: Financial Markets and Changing Climate Risks. Fondazione Eni Enrico Mattei, Milano.

Clark, W., Jaeger, J., Eijndhoven, J.V., 2001. Managing Global Environmental Change. The Social Learning Group, Learning to Manage Global Environmental Risks, vol. 1. The MIT Press Cambridge, London.

Cogan, D., 2003. Corporate governance and climate change: making the connections. CERES Sustainable Governance Project Report Available at: http://www.ceres.org/pdf/ceres_cg_rprt.pdf, 15.07.2003.

Cogan, D., 2004. Investor Guide to Climate Risk, Action Plan and Resource for Plan Sponsors Fund Managers and Corporations. A Publication of the Investor Network on Climate Risk. CERES, Boston.

Commission of European Communities, 2005. Proposal for a Council Directive on passenger car related taxes. COM(2005) 261 final, Brussels.

Commission of the German Corporate Governance Code, 2005. German Corporate Governance Code, 6. Transparency. Available at: http://www.corporate-governance-code.de/eng/kodex/6.html, 10.12.2005.

Dobbs, R., Manson, N., Nyquist S., 2006. Capital discipline for Big Oil. McKinsey on Finance, pp. 6–11.

European Environment Agency, 2004. Impacts of Europe's Changing Climate, an Indicator-Based Assessment. EEA Report, vol. 2/2004. Copenhagen.

Faff, R., Brailsford, T., 1999. A Test of a Two-Factor Market and Oil Pricing Model. Working Paper, vol. 99-05. The Australian National University, Melbourne.

Federal Institute for Geosciences and Natural Resources, 2005. Reserves, resources and availability of energy resources 2004. Available at: http://www.bgr.bund.de, 01.03.2006.

Fishhaut, E., 2003. Petrochemical firms take a stand against oil price volatility. Market focus May 2003. Available at: http://eprm.com, 10.11.2005.

Garz, H., Volk, C., 2003. Carbonomics – Value at Risk Through Climate Change. WestLB AG, London, Duesseldorf.

Garz, H., Kudszus, V., Volk, C., 2004. Insurance and Sustainability – Playing with Fire. WestLB AG, London.

Gerlagh, R., Lise, W., 2005. Carbon taxes: a drop in the ocean, or a drop that erodes the stone? The effect of carbon taxes on technological change. Ecological Economics 54, 241–260.

Goldman Sachs, 2006. The top ten financial risks to the global economy: a dialogue of critical perspectives. Global Markets Institute at Goldman Sachs and Co. Available at: http://www.gs.com, 05.03.2006.

Henderson Global Investors, 2002. Socially Responsible Investments. Climate Change Position Paper, London.

Hirsch, R., Bezdek, R., Wendling, R., 2005. Peaking of World Oil Production: Impacts, Mitigation, and Risk Management. SAIC, MISI.

Hoffmann, V.H., Trautmann, T., in press. The role of industry and uncertainty in regulatory pressure and environmental strategy. Academy of Management Best Conference Paper 2006 ONE.

Hotelling, H., 1931. The economics of exhaustible resources. Journal of Political Economy 39, 137–175.

Hubbert, M., 1974. M. King Hubert on the nature of growth. National Energy Conservation Act of 1974. Hearings before the Subcommittee on the Environment of the Committee on the Interior and Insular House of Representatives.

Innovest Strategic Value Advisors, 2002a. Climate change and the financial services industry. Module 1 – threats and opportunities. Prepared for the UNEP Finance Initiatives Climate Change Working Group. Available at: http://www.unepfi.net/cc/mod1_ccwg_unepfi.pdf, 02.07.2003.

Innovest Strategic Value Advisors, 2002b. Value at risk: climate change and the future of governance. Prepared for CERES

Sustainable Governance Project Report. Available at: http://www.innovestgroup.com/pdfs/climate.pdf, 07.07.2003.

Innovest Strategic Value Advisors, 2003a. The carbon value at risk portfolio audit. Available at: http://www.innovestgroup.com, 15.01.2004.

Innovest Strategic Value Advisors, 2003b. Carbon finance and the global equity markets. Carbon Disclosure Project. Available at: http://www.cdproject.net, 10.12.2005.

Innovest Strategic Value Advisors, 2004. Climate change and shareholder value in 2004. Carbon Disclosure Project. Available at: http://www.cdproject.net, 10.12.2005.

Innovest Strategic Value Advisors, 2005a. Carbon Disclosure Project 2005, on behalf of 155 investors with assets of $21 trillion. Available at: http://www.cdproject.net, 10.12.2005.

Innovest Strategic Value Advisors, 2005b. Analyzing carbon risk and financial out-performance potential. Available at: http://www.innovestgroup.com/pdfs/CDP_Analysis.pdf, 05.12.2005.

Innovest Strategic Value Advisors, WWF, 2003. Power switch: impacts of climate policy on the global power sector. Available at: http://www.innovestgroup.com/pdfs/2003-11-PowerSwitch.pdf, 05.12.2005.

IEA, 2004. Oil Crises and Climate Challenges, 30 Years of Energy Use in IEA Countries. International Energy Agency, Paris.

IEA, 2005a. Oil Market Report, a Monthly Oil Market and Stocks Assessment. International Energy Agency, Paris. October.

IEA, 2005b. World Energy Outlook 2005. International Energy Agency, Paris.

Intergovernmental Panel on Climate Change, 2001. Climate Change 2001. Third Assessment Report. Cambridge University Press, Cambridge, New York.

Joshi, S., Khanna, M., Sidique, S., 2005. Effect of Environmental Management Systems on Investor Reactions to Emission Information. Academy of Management Best Conference Paper 2005 ONE: D1–D6.

Kiernan, M., 2002. Taking Control of Climate. Financial Times. 25.11.2002.

King, A., Lenox, M., 2001. Does it really pay to be green? An empirical study of firm environmental and financial performance. Journal of Industrial Ecology 5 (1), 105–116.

Kolk, A., Levy, D., 2001. Winds of change: corporate strategy, climate change and oil multinationals. European Management Journal 19 (5), 501–509.

Kolk, A., Pinske, J., 2004. Market strategies for climate change. European Management Journal 22 (3), 304–314.

Kolk, A., Pinske, J., 2005. Business responses to climate change: identifying emergent strategies. California Management Review 47 (3), 6–20.

Kunreuther, H., 1978. Disaster Insurance Protection. New York.

Lehman Brothers Equity Research, 2002. Oil and gas: E and P (Large Cap) – industry update. Available at: http://www.climatechangecentral.com/news_room/SC02_Kyoto.pdf, 24.07.2003.

Lovins, A., Datta, E., Bustners, O.E., Koomery, J., Glasgow, N., 2005. Winning the Oil Endgame – Innovation for Profits, Jobs, and Security. Rocky Mountains Institute, Snowmass, Colorado.

Lund, P., 2003. Emissions trading: carbon will become a taxing issue for European utilities. Standard and Poor's publication. Available at: http://www2.standardandpoors.com, 24.07.2003.

Mansley, M., 2002. Risking shareholder value? ExxonMobil and climate change, an investigation of unnecessary risks and missed opportunities. A Claros discussion paper. Available at: http://www.campaignexxonmobil.org/pdf/RiskingValue.pdf, 20.12.2005.

Mansley, M., 2003. Sleeping tiger, hidden liabilities: amid growing risk and industry movement on climate change, ExxonMobil falls farther behind. A Claros discussion paper. Available at: http://www.ceres.org, 02.07.2003.

Mansley, M., Dlugolecki, A., 2001. Climate Change – a Risk Management Challenge for Institutional Investors. Universities Superannuation Scheme, Discussion Paper, vol. 1. London.

Margolis, J., Walsh, J., 2001. People and Profits? The Search for a Link Between a Company's Social and Financial Performance. Lawrence Erlbaum Associates, Publishers, Mahwah, New Jersey.

McDonough, W., Braungart, M., 2002. Cradle to Cradle: Remaking the Way We Make Things. North Point Press, New York.

Meacher, M., 2004. Plan Now for a World Without Oil. Financial Times. 05.01.2004.

Merrill Lynch, 2005. Energy Security and Climate Change, Investing in the Clean Car Revolution. World Resource Institute.

Mills, E., 2005. Insurance in a climate of change. Science 309, 1040–1044.

Morgan, M., Henrion, M., 1995. Uncertainty, a Guide to Dealing with Uncertainty in Qualitative Risk and Policy Analysis. Cambridge University Press, Cambridge.

Munich Re, 2005. Topics Geo, Annual review: Natural Catastrophes 2004. Munich Reinsurance Group, Munich.

Murphy, C., 2002. The profitable correlation between environmental and financial performance: a review of the research. Available at: http://www.lightgreen.com/files/pc.pdf, 01.12.2005.

Nordhaus, W.D., 1994. Managing the Global Commons: The Economics of Climate Change. MIT Press, Cambridge.

Odell, P., 2004. Why Carbon Fuels Will Dominate the 21st Century's Global Energy Economy. Multi-Science Publishing, Brentwood, Essex.

Orlitzky, M., Schmidt, F., Rynes, S., 2003. Corporate social and financial performance: a meta-analysis. Organization Studies 24 (3), 403–441.

Powell, M., Langley, R., McVicar, M., Copp-Barton, J., 2003. Aviation Emissions, Another Cost to Bear. Dresdner Kleinwort Wasserstein Research, London.

Reinhaud, J., 2003. Emission Trading and its Possible Impacts on Investment Decisions in the Power Sector. International Energy Agency Information Paper, Paris.

Reinhaud, J., 2005. Industrial Competitiveness Under the European Union Emission Trading scheme. International Energy Agency Information Paper, Paris.

Repetto, R., Austin, D., 2000. Pure Profit: the Financial Implications of Environmental Performance. World Resource Institute, Washington, DC.

Repetto, R., Henderson, J., 2003. Environmental Exposures in the U.S. Electricity Utility Industry. Yale School of Forestry and Environmental Studies, New Haven.

Reynolds, D., 1999. The mineral economy: how prices and costs can falsely signal decreasing scarcity. Ecological Economics 31, 155–166.

Roberts, P., 2004. Running Out of Oil – and Time. Los Angeles Times. 07.03.2004.

Ruth, M., 2006. A quest for the economics of sustainability and the sustainability of economics. Ecological Economics 56, 332–342.

Ruth, M., Davidsdottir, B., Laitner, S., 2000. Impacts of market-based climate change policies on the US pulp and paper industry. Energy Policy 28, 259–270.

Ruth, M., Davidsdottir, B., Amato, A., 2004. Climate change policies and capital vintage effects: the cases of US pulp and paper, iron and steel, and ethylene. Journal of Environmental Management 70, 235–252.

Scott, A., Pearse, P., 1992. Natural resources in a high-tech economy: scarcity versus resourcefulness. Resources Policy 18 (3), 154–166.

Sadorsky, P., 1999. Oil price shocks and stock market activity. Energy Economics 21, 449–469.

Solow, R., 1974. The economics of resources or the resources of economics. American Economic Review 66, 1–114.

Stanwick, P., Stanwick, S., 1998. The relationship between corporate social performance and organizational size, financial performance, and environmental performance: an empirical examination. Journal of Business Ethics 17 (2), 195–204.

Stern Review, 2006. What is the economics of climate change? Discussion paper 31 January 2006. Available at: http://www.sternreview.org.uk, 05.03.2006.

Swiss Re, 2005. Sigma, Natural Catastrophes and Man-Made Disasters in 2004. Swiss Reinsurance Company, Zurich.

The Center for Health and the Global Environment, 2005. Climate change futures – health, ecological and economic dimensions. Available at: http://www.climatechangefutures.org/pdf/CCF_Report_Final_10.27.pdf, 05.12.2005.

The Danish Board of Technology, 2003. Oil-Based Technology and Economy – Prospects for the Future. Preliminary Edition, Copenhagen Conference on Oil Demand Production and Cost, Copenhagen.

The Economist, 2005a. Oil in Troubled Waters. A Survey of Oil, April 30th 2005.

The Economist, 2005b. Consider the Alternatives. April 28th 2005.

Tierney, J., 1990. Betting the Planet. New York Times Magazine December 2, 52 (Section 6).

Tol, R., 1998. Climate change and insurance a critical appraisal. Energy Policy 26 (3), 257–262.

Tol, R., 2002a. Estimates of the damage costs of climate change. Part I. Benchmark estimates. Environmental and Resource Economics 21, 47–73.

Tol, R., 2002b. Estimates of the damage costs of climate change. Part II. Dynamic estimates. Environmental and Resource Economics 21, 135–160.

UNEP Finance Initiative, 2003. 0.618 ... The golden ratio. Issue 3. Available at: http://www.unepfi.org, 20.12.2005.

Union of Concerned Scientists 2002. Energy Security, Solutions to Protect America's Power Supply and Reduce Oil Dependence. Cambridge.

Vellema, S., Tuil, R., Eggink, G., 2003. Sustainability, agro-resources and technology in the polymer industry. In: Steinbüchel, A. (Ed.), Biopolymers. WILEY-VCH, Weinheim, pp. 339–363.

Vogel, D., 2005. The Market for Virtue, the Potential and Limits of Corporate Social Responsibility. Brookings Institution Press, Washington D.C.

Walker, W., 2000. Entrapment in large technology systems: institutional commitment and power relations. Research Policy 29, 833–846.

Wellington, F., Sauer, S., 2005. Framing Climate Risk in Portfolio Management. Ceres, World Resource Institute, Boston, Washington, D.C.

Wellington, F., Sauer, S., Fox C., Gardiner, D., 2004. Questions and Answers for Investors on Climate Risk. Ceres, World Resource Institute, Boston, Washington, D.C.

Wolfson, D., Koopmans, C., 1996. Regulatory taxation of fossil fuels: theory and policy. Ecological Economics 19, 55–65.

World Economic Forum, 2000. Business leaders say climate change is our greatest challenge. Available at: http://www.weforum.org/site/knowledgenavigator.nsf/Content/_S15317 open, 20.12 05.

World Resource Institute, 2002. Changing oil: emerging environmental risks and shareholder value in the oil and gas industry. Available at: http://business.wri.org/project_description2.cfm?ProjectID=32, 03.07.2003.

Annex

Paper V

Ecology-Driven Real Options: An Investment Framework for Incorporating Uncertainties in the Context of the Natural Environment

Timo Busch
Volker H. Hoffmann

ABSTRACT. The role of uncertainty within an organization's environment features prominently in the business ethics and management literature, but how corporate investment decisions should proceed in the face of uncertainties relating to the natural environment is less discussed. From the perspective of ecological economics, the salience of ecology-induced issues challenges management to address new types of uncertainties. These pertain to constraints within the natural environment as well as to institutional action aimed at conserving the natural environment. We derive six areas of ecology-induced uncertainties and propose ecology-driven real options as a conceptual approach for systematically incorporating these uncertainties into strategic management. We combine our results in an integrative investment framework and illustrate its application with the case of carbon constraints.

KEY WORDS: ecological economics, uncertainty, natural environment, real options, investment planning

Introduction

In business ethics literature, there is a comprehensive debate of the role, extent, and necessity of ethical decision making in business (e.g., Donaldson and Dunfee, 1999; Jones, 1991; Knouse and Giacalone, 1992; Trevino, 1986). Ethical decision making in organizations is impacted by several content variables, namely individual variables, the job context, the organizational context, and the external environment[1] (McDevitt et al., 2007). With respect to the latter, external forces such as societal expectations, political institutions, or industry norms "can create environmental uncertainty" (McDevitt et al., 2007, p. 222) that is important to be addressed within internal management decisions. In this article, we investigate the role of emerging environmental uncertainties in such decisions. More specifically, we focus on uncertainties stemming from the natural environment for two reasons: first, ecological issues have not been of special emphasis in early work in business ethics (e.g., Garrett, 1966; Sharp and Fox, 1937) but are now of particular interest within the business ethics literature (e.g., Crane and Matten, 2007; Ferrell et al., 2002; Lawrence et al., 2005). Second, the issue of environmental uncertainty and its effects on organizations are very prominent in management literature (see Buchko, 1994; Miller and Shamsie, 1999 for reviews). However, what kinds of business-relevant uncertainties stem from the natural environment, what sort of risks these uncertainties pose for firms, and how these risks can be dealt with within strategic decision making, especially in corporate investment planning, are areas in which the discussion has been limited.[2]

We derive our main arguments from the literature on ecological economics and elucidate that new business-relevant issues stem from constraints within the natural environment and from institutional action aimed at conserving the natural environment. We refer to these issues as ecology-induced issues and suggest that they change the business environment and, consequently, constitute a new challenge for strategic management (Shrivastava, 1995). More specifically, we argue that strategic management has to attend to uncertainties stemming from these ecology-induced issues. We elucidate six areas of ecology-induced uncertainties and conclude that flexible adjustments in investment strategies as a response to these uncertainties are important. However, traditional financial assessment methods such as net present value (NPV) calculations fall short of capturing the value of these adjustments. In a similar context, Husted (2005) suggests that real options help to alleviate this shortcoming and provide a basis to better understand the strategic relevance of corporate social responsibility (CSR). He argues that real options are able to reflect managerial risks, and the value of such an option "increases as perceived environmental uncertainty increases" (Husted, 2005, p. 180). As such, real options facilitate strategic decisions in situations when a precise assessment of an investment's profitability is limited due to an uncertain business environment and when management is prompted to consider flexible adjustments to its original strategic plans.

We build on Husted's approach and investigate perceived uncertainties in the context of the natural environment. We take an instrumental perspective (Friedman, 1962; Garriga and Mele, 2004) by focusing on corporate profitability and suggest ecology-driven real options as a conceptual approach for investment planning under ecology-induced uncertainties. Based on this, we delineate an integrative investment framework and apply it to the case of carbon constraints.

Ecological economics and salience of ecology-induced issues

Firms are increasingly confronted with ethical decisions on the relationships formed between business and society (De Tienne and Lewis, 2005). Especially in the context of the natural environment, scholars have found that new issues emerge (e.g., Bansal and Roth, 2000) which constitute business-relevant topics (e.g., Starik, 1995) and can be interpreted as "catalyst for a new round of creative destruction" (Hart and Milstein, 1999). In this sense, the business environment in general goes through a transitional phase (Porter and van der Linde, 1995), and previous views that took certain business conditions for granted have to be challenged (Gladwin et al., 1995; King, 1995).

The interactions of a firm with the natural environment are discussed in two major streams of neoclassical economic literature: natural resource economics mainly considers the economy's extraction of resources from the natural environment, while environmental economics puts emphasis on the economy's material flows into the natural environment (Common and Stagl, 2005). The concept of ecological economics seeks to address both literature streams (Costanza et al., 1991) and emphasizes the role and impact of human activities with respect to the natural environment. Thus, ecological economics can be defined as "the study of the human economy as part of nature's economy" (Common and Stagl, 2005, p. 16). We base our initial arguments on this perspective and observe the salience of ecology-induced issues in two dimensions that influence the corporate business environment: constraints within the natural environment and institutional action aimed at conserving the natural environment.

In the first dimension, we refer to ecology-induced issues as constraints within the natural environment, which relate to the earth's endowment with natural resources and the carrying capacity of the natural environment. The natural resource endowment is determined by ecological limitations such as the finite reserves of natural resources within the ecosphere and the depletion of these resources. The carrying capacity addresses the ability of the ecosystem to absorb pollution discharges such as air emissions and delimits the critical flows of these substances from the anthroposphere to the ecosphere. Both the endowment of natural resources and the carrying capacity of the natural environment are normally considered stable business conditions, i.e., firms take a technocentric view and presume that the current status quo obtains within a given planning horizon (Gladwin et al., 1995). However, taking an ecological economics' perspective, the business conditions under which firms operate are increasingly changing under the growing impact of these ecology-induced issues.

In the second dimension, we refer to ecology-induced issues as institutional action aimed at conserving the natural environment. In general, institutional action refers to activities that become institutionalized over time (Scott, 2001) and tend to be enduring without further justification, socially accepted, and resistant to change (Oliver, 1992). In our context, we consider the interorganizational level of institutional theory (Oliver, 1997) and focus on "both formal and informal pressures exerted on organizations by other organizations upon which they are dependent and by cultural expectations in the society within which organizations function" (DiMaggio and Powell, 1983, p. 150). As such, we frame institutional action as human interventions and responses meant to protect the integrity of the natural environment. On the one hand, this action describes the contribution of governments, environmental activists, journalists, or scientists (Hannigan, 2006) to a changing institutional environment. On the other hand, informal social movements, interpreted as loosely organized collective actions, also shape the institutional environment (Benford and Snow, 2000; Polletta and Jasper, 2001). In short, the discussed institutional action is starting to fundamentally alter the institutional environment and, hence, the way business works within society. Therefore conformity to such social expectations is important for firms as it "contributes to organizational success and survival" (Oliver, 1997, p. 699).

Uncertainties in the context of ecology-induced issues

In both dimensions, the ecology-induced issues constitute a new and salient driver of a changing business environment. However, they do not emerge in a continuous and predictable manner. Instead, the future availability of resources and the ecosystem's dynamics are uncertain (Chichilnisky and Heal, 1998; Heal, 1998; King, 1995), as is the process of interaction involved in institutional change (Lepoutre et al., 2007; McDevitt et al., 2007). Since environmental changes or corresponding uncertainties stimulate changes within organizations (Damanpour and Evan, 1984), firms need to specifically address emerging ecology-induced uncertainties. However, we argue that these uncertainties are in some respects fundamentally new and different compared to other uncertainties within the business environment, as they pertain to conditions that firms have taken for granted as enduring and stable. We see three interrelated reasons for this.

First, forecasting conditions in the natural environment such as ecological limitations or ecosystems' thresholds over the long term is difficult (King, 1995), as methodologies to appropriately deal with "external shocks, non-linear responses, and discontinuous behavior" (Clark, 1986, p. 31) are scarce and sufficient ex post data or long-term time series are often not available. These are needed for reliable predictions of future developments within the natural environment and related uncertainties. Similarly, institutional processes may appear stable for a certain time when in fact organizational fields and institutions co-evolve (Hoffman, 1999) and thus are rather not static (Greenwood et al., 2002). For example, stakeholder pressures on firms have increased dramatically (Dawkins and Lewis, 2003), and it is hard to predict how stakeholders' expectations and claims will develop in the future.

Second, understanding these uncertainties in a managerial context is difficult because of the problem of chaos and complexity (Clark, 1986; Prigogine and Stengers, 1984; Wheatley, 1999). It is a rule of ecology that everything is interconnected and each environmental insult will likely redound on society (King, 1995). Therefore, concerns about the global ecosystem "lead to the generation of crude and difficult-to-operationalize axioms" (Gladwin et al., 1995, p. 891). From an institutional point of view, the direction and pace of changes in the business environment vary across and within institutional sectors (Greenwood and Hinings, 1996). This offers strategic management a palette of response options but does not guarantee that any of them will meet societal expectations.

Third, the change in business conditions due to ecology-induced uncertainties can be rapid and massive. The overshoot of sustainable limits could cause a sudden environmental collapse (Meadows et al., 1972), which in organizational theory has been termed ecological surprise (King, 1995). Furthermore, natural disasters are not necessarily static, isolated phenomena (Hannigan, 2006) but constitute a threat of "massive discontinuous ecological changes" (Winn and Kirchgeorg, 2005, p. 233). From an institutional point of view, disruptive events such as the Rio Summit, catastrophes such as the Exxon Valdez oil spill, and legal or administrative activities like the release of environmental white papers (Hannigan, 2006) can result in sharp institutional changes (Hoffman, 1999).

Perception of ecology-induced uncertainties

It is a new challenge for firms to discard their prior 'taking-for-granted' view in the context of the natural environment and to analyze the relevance of corresponding uncertainties for strategic decisions. In this respect, a distinction can be made between perceived and objective uncertainties (Aragon-Correa and Sharma, 2003; Boyd et al., 1993). We build on the literature of perceived uncertainties for two reasons: first, knowledge, a precondition for assessing the future business environment, is determined more by perception than by objectivity (Hambrick et al., 2005; Smircich and Stubbart, 1985). Second, also ethical decision making stems from subjective (i.e., perception-based) assessments, with respect to both its behavioral choice component and its normative-effective component (Trevino and Youngblood, 1990).

Milliken (1987) defines uncertainty as an individual's perceived inability to predict a future condition accurately; such uncertainties pertain to general external events, cause–effect relationships between a firm and its environment, and management decision outcomes (Miller and Shamsie, 1999). Three types of uncertainty can be distinguished: environmental state uncertainty refers to the inability to forecast future industry or market developments. This results from conditions in the business environment or one of its components that all firms face, such as demand volatility, price increases, or regulatory pressure. Organizational effect uncertainty describes the inability to predict the impact of environmental events or changes on firms. It derives from a lack of knowledge and skills to understand the cause–effect relationship between environmental effects or changes and the individual corporate exposure. Decision response uncertainty represents the lack of knowledge concerning suitable response options and/or the inability to anticipate the consequences of individual decisions. We relate these three types of uncertainty to the previously discussed two dimensions of ecology-induced issues. As a consequence, we derive six areas of ecology-induced uncertainties that matter for corporate responsiveness to an ecology-induced change in the business environment (Table I).

With respect to constraints within the natural environment, management faces environmental state uncertainty regarding the general *extent and timing of ecological limitations* and their influence on the corporate business environment. Beyond that, management faces organizational effect uncertainty regarding the *magnitude and direction* of the influence of such ecological limitations: individual firms exhibit different exposures to ecology-induced constraints due to their unique position in industries, value chains, and geographies and resources, and capabilities to cope with these constraints are firm-specific. Anticipating likely constraints, firms are prompted to alter their strategy, notably in terms of investment decisions. To choose a successful response strategy, several firm-specific circumstances such as the general availability of technical alternatives have to be taken into account. Therefore, management faces decision response uncertainty regarding the firm's *own alternatives* for adequately reacting to and foreseeing the *consequences of coping* with ecology-induced constraints.

In light of institutional action aimed at conserving the natural environment, management faces

TABLE I
Six areas of ecology-induced uncertainties

Change of business environment due to	Environmental state uncertainty pertains to	Organizational effect uncertainty pertains to	Decision response uncertainty pertains to
Constraints within the natural environment	Extent and timing of ecological limitations	Magnitude and direction of ecology-induced constraints for the firm	Own alternatives for and consequences of coping with ecology-induced constraints
Institutional action aimed at conserving the natural environment	Scale and scope of human responses to ecological issues	Exposure to and relevance of ecology-induced institutional changes for the firm	Own alternatives for and consequences of adjusting to ecology-induced institutional changes

environmental state uncertainty vis-à-vis the general *scale and scope of human responses to* ecological issues and their influence on the corporate business environment. Moreover, the extent to which organizations exert formal and informal pressures and how social movements will affect individual firms is uncertain. Therefore, management faces organizational effect uncertainty regarding the *exposure to and relevance of* ecology-induced institutional changes. To choose a successful response strategy, several firm-specific characteristics are important, such as the firm's ability to reliably fulfill stakeholders' requirements and expectations. Therefore, management faces decision response uncertainty regarding the firm's *own alternatives* for adequately reacting to and foreseeing the *consequences of adjusting* to ecology-induced institutional changes.

In order to facilitate addressing these uncertainties in the managerial context, management has to understand the interplay between them. On the one hand, there is no primacy or dependence among the six areas: if a change in the business environment can be detected due to one specific issue, each of the areas can represent an independent source of uncertainty for a firm. The relevance of each uncertainty depends on the individual perception by management and the general business circumstances, such as the competitive landscape or the firm's technological possibilities. On the other hand, the six areas are interrelated over time: once management has solved decision response uncertainty by taking a certain action, this in turn feeds back to environmental state and organizational effect uncertainties. As such, some of the state and effect uncertainties might even be created by corporate responses to other ecology-induced issues.

Determining the profitability of investments

Generally, the anticipation of future developments in the business environment and their integration into the assessment of future cash flows constitute an important part of a firm's strategic management (Thompson, 1967). However, if these assessments face an uncertain business environment, an appropriate consideration of uncertainty is important for successful investment planning (Dixit and Pindyck, 1994). The economic consequence of a firm's exposure to uncertainties in the business environment is financial risk (Amram and Kulatilaka, 1999). For a comprehensive analysis of this risk in the context of ecology-induced issues, the six derived areas of uncertainties facilitate assessing the future profitability of investments. Standard methods for investment appraisal such as NPV calculations estimate and discount future cash flows but face two important limitations.

First, emerging ecology-induced issues may prompt management to alter its original plans. Consequently, firms need to reflect on different possible developments of the business environment and to determine the value of flexible adjustments in their investment strategy at a later stage, e.g., the value of switching, extending, or stopping the investment. However, the possibility of changing the investment strategy is not built into typical financial NPV models as they usually consider only one likely return stream of an entire project (Trigeorgis, 1988). Using NPV as an investment criterion is thus most suitable for firms operating in a fairly stable business environment.

Second, NPV logic usually applies a higher discount rate when returns appear to be more uncertain (Baecker and Hommel, 2004). As a result, the present value of the free cash flows decreases. Therefore, this valuation method captures the possibility that actual returns might be lower than expected, but the possibility that actual returns might be higher is not appropriately reflected in the valuation process (Cornelius et al., 2005). Hence, this inherent cognitive bias and risk-averse perspective of considering only possible negative effects might prompt managers to reject a project solely on the basis of a high level of uncertainty, thereby neglecting the opportunity perspective of ecology-induced investments.

The central challenge of a farsighted management is to adequately respond to unforeseen changes by incorporating flexibility into investment appraisals (Copeland and Antikarov, 2001). Available methods include Monte Carlo simulation (e.g., Clemen, 1996), dynamic programming (e.g., Cormen et al., 2001), and real options theory (e.g., Black and Scholes, 1973). For incorporating flexibility as a response to ecology-induced uncertainties, this article focuses on the last. In general, real options theory assumes an initial investment is to be made (Dixit and Pindyck, 1995); management must then decide whether to harvest or cultivate the initial investment (Adner and Levinthal, 2004). As such, "real options can help decision makers assess the profitability of new projects and understand whether and when to proceed with the later phases of projects that have already been initiated. [...] Real options are especially valuable for projects that involve both a high level of uncertainty and opportunities to dispel it as new information becomes available" (Copeland and Keenan, 1998, pp. 129–130). Following this logic, the result is an extended NPV, which consists of the standard NPV (without considering the value of flexibility) and the option value (Trigeorgis, 1995). The latter describes the value of flexible adjustments of the investment strategy.

When management intends to utilize real options, five parameters have to be determined: the present value of an investment's operating assets, the expenditure required to acquire the investment's assets, the considered time length, the time value of money, and underlying volatilities (Luehrman, 1998). Summarizing those parameters, Dixit and Pindyck (1994) name three determinants for incorporating flexibility into the process of assessing investments' profitability in an uncertain business environment: (a) analysis of the underlying investment conditions: what are the parameters of the project and what is the value of the investment under current conditions? (b) appraisal of volatilities: what is the likely distribution of future revenues? (c) assignment of a time to invest: when is the best time to invest?

In order to determine the option value in the context of ecology-induced issues, the orresponding uncertainties have to be translated into probabilities, i.e., interpreted as corporate risks. As one important assumption, the literature on real options differentiates between public and private risks (Borison, 2005), both in theoretical work (e.g., Dixit and Pindyck, 1994; Trigeorgis, 1988) and in managerial applications (Amram and Kulatilaka, 1999; Luehrman, 1998). Public risks are market risks, i.e., estimates for such risks can be observed on the market in alternative portfolios. These rather pertain to environmental state and organizational effect uncertainties. Private risks are firm-specific and require subjective estimates. As such, they can relate to all ecology-induced uncertainties in a similar manner. Both types of risk have to be considered when analyzing the three determinants of investments' profitability. As a result, management can consider the five types of real options delineated in Table II for practical application within investment decisions (Amram and Kulatilaka, 1999).

Ecology-driven real options

In the following, we exemplify the application of these five types of real options with a fictitious company which considers investing in a new environmentally sound production system. In order to develop the underlying technology, the company has already made R&D investments which represent an initial investment. The company now reflects on different ecology-induced uncertainties that could affect the future profitability of the production system once it is installed. In this situation, the application of ecology-driven real options is suitable: an option to defer could, for example, be built into the purchasing agreements of major components to allow postponing the start of the production system in case it turns out to be unprofitable under current market conditions. Conversely, if the new technology exceeds profitability forecasts a growth option can be realized to generate additional revenues through the increased sale of environmentally improved products. An option to extend could be created if the company is able to transfer the technology into related production systems in different fields, for example, with modularized technological components. Furthermore, the technology could be designed in such a way that an option to switch between different types of the environmentally sound system allows adjusting to changes in market conditions. Finally, management is able to create an option to abandon if some value of the production system can be retained even if its operation is discontinued at a later point in time. This can be the case, for example, if gained R&D insights can be used within other projects.

In light of the potential advantages outlined above, several scholars have started to use real options theory in order to address specific questions related to the natural environment. Table III illustrates some prominent examples. Most of these studies discuss the application of real options theory

Ecology-Driven Real Options

TABLE II
Five types of real options within investment decisions

Types of option	Management flexibility	Description
Option to defer	Deferring the exercise date into the future	An option to defer allows the management to postpone the start of an investment. This applies to investments that are not profitable under current conditions but might become profitable at a later stage
Option to grow	Flexible adjustment of project's scope	Growth options can be adequate in situations where an initial investment turns out to be profitable. While building on this investment, further investments generate additional revenues at a later stage
Option to extend	Broadening the utilization of gained knowledge	Considering options to extend, firms are able to utilize an initial investment in related areas afterward if the conditions are favorable. Management is able to transfer technologies or knowledge gained to other projects
Option to switch	Flexible choice of path	Within a project's lifetime, management may have the option to move back and forth between different possibilities to utilize the initial investment, depending on each possibility's profitability
Option to abandon	Stop project	An option to abandon describes the possibility to stop a project at a later stage while retaining the ability to capture a remaining value of the initial investment. A reason for stopping a project could be a change in market conditions

Source: Extended from Amram and Kulatilaka (1999).

TABLE III
Applications of real option theory in the context of the natural environment

Authors	Areas of application	Types of real option
Blyth et al. (2007)	Investment decisions in the power sector under uncertain climate policy	Option to defer
Cortazar et al. (1998)	Investments in environmental technologies under varying output price levels	Option to defer, option to extend, and option to abandon
Laurikka and Koljonen (2006)	Investment decisions in the power sector in face of the EU ETS and fossil fuel prices	Option to defer, option to switch
Lin et al. (2007)	Timing of environmental pollution policy	Option to defer
Yang et al. (2008)	Investment decisions in the power sector in face of regulatory uncertainty	Option to defer

in the context of investment decisions in the power sector and consider the option to defer. A frequent conclusion is that due to uncertainties in climate policy management should pursue a waiting strategy and postpone investments.

Extending the basic idea of these studies, we devise a more general argument: on the one hand, we discussed that the salience of ecology-induced issues increases within the business environment. On the other hand, the underlying uncertainties exhibit specific characteristics which are not adequately treated in strategic management. Therefore, we propose a real options logic as a constitutive element of investment decisions in which managerial flexibility toward the derived six areas of ecology-induced uncertainties is required. This is of special relevance when assessing the profitability of investments with long lead and amortization times. With such a real options logic, management can reduce what Rugman and Verbeke (1998) call irreversible green mistakes: starting from a competitive advantage point of view, Rugman and Verbeke (1998) explain the importance of analyzing the flexibility of resource commitments as well as their leveraging potential for improving a firm's performance with respect to the natural environment. In the logic of their framework, the greater the flexibility and leveraging potential of resource commitments, the higher the utility of applying real options. In such business situations, it can be reasonable to consider one or several of the five types of real options mentioned above. Which option is most suitable depends on the characteristics of the required resource commitments (e.g., interchangeability of required resources), the firm-specific risk exposure (e.g., competitive landscape), the individual context of the investment decision (e.g., long-term amortization periods, path dependencies), the salience of ecology-induced issues, and the corresponding perception of uncertainties.

Based on these ideas, we derive an integrative investment framework that facilitates the incorporation of ecology-driven real options in investment decisions (Figure 1). This framework combines four steps under one conceptual umbrella. In the first step, ecology-induced changes in the business environment are analyzed that stem from constraints within the natural environment and related institutional action. In the second step, it is evaluated how these changes are likely to influence the general state of the business environment, to affect the organization, and to be relevant for decisions regarding how

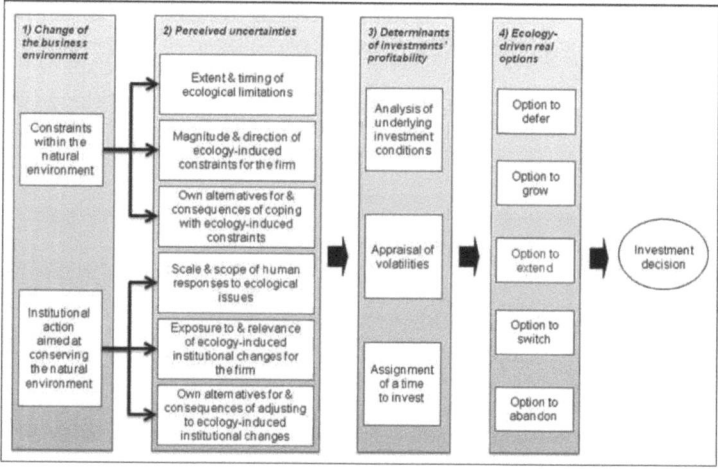

Figure 1. Integrative investment framework for ecology-driven real options.

to respond to these changes. Accordingly, one or more of the six areas of ecology-induced uncertainties might turn out to be important to focus on within an investment's planning process. In the third step, the consequences of these uncertainties for the determinants of investment profitability are assessed to offer a differentiated reflection from a risk perspective. In the fourth step, based on the effect of uncertainties on the determinants, one or more of the five real options can be considered. In this way, investment decisions in the face of ecology-induced uncertainties are facilitated.

Example: carbon constraints

We apply the investment framework by referring to carbon constraints. These relate on the one hand to the disposition of fossil fuels and on the other hand to direct and indirect climate change effects (Busch and Hoffmann, 2007). Carbon constraints can be considered a very prominent example for ecology-induced issues and uncertainties in the 21st century. As almost every industry is dependent on utilizing energy or fossil fuels, the steps below facilitate investment planning for firms independently of their industry affiliation. In the last step, we illustrate how individual firms have already implemented ecology-driven real-options-thinking.

Step 1: analyzing changes of the business environment

The availability of fossil fuels can be considered an emerging ecology-induced issue for firms: for example, rising oil prices and increasing price volatilities may be a first sign that markets are reacting to imminent concerns over the earth's endowment of fossil fuels (Hirsch et al., 2005). This *constraint within the natural environment* has macroeconomic implications for economies (Kuik, 2003) and, thus, influences the business environment. Furthermore, empirical data and findings of the IPCC (2007) prove that climate change has emerged as an ecology-induced issue, as greenhouse gas emissions are on the verge of exceeding the carrying capacity of the natural environment. Negative economic effects through adaptation and mitigation measures can be expected (Stern, 2006), which, in turn, will also change the business environment. Furthermore, climate change-related extreme weather events can destroy thriving business environments (Schwartz, 2007).

With respect to *institutional action*, a prevailing carbon constraint for firms is the European Emission Trading Scheme (EU ETS) that was launched in 2005.[3] The EU assigns amounts of CO_2 emission allowances to their member states which, in turn, grant these allowances to firms according to so-called National Allocation Plans. Firms can utilize them to fulfill their obligation to render allowances for their CO_2 emissions or trade them on the carbon market. From January to March 2008, the daily trading volumes centered around 10 million tonnes of CO_2 with an average price of 21 EUR per tonne of CO_2 (Point Carbon, 2008). Hence, changes in the natural environment lead to the implementation of this new regulatory framework, which constitutes an institutional change that influences the business environment (Hourcade et al., 2007).

Step 2: evaluating the perception of corresponding uncertainties

These previously described developments relate to all six areas of ecology-induced uncertainties. For example, firms face environmental state uncertainty regarding the future *extent and timing* of oil availability. This depends on a large number of factors, e.g., the discovery of new resource fields. Furthermore, price increases of crude oil affect corporate stock returns (Sadorsky 1999), but the *magnitude and direction* of oil scarcity are difficult to anticipate, i.e., the effects for individual firms are subject to uncertainty. Some industries are dependent on utilizing oil and might bear price increase as long as they are able to pass them on to customers. However, it is uncertain when and to what extent disruptive technologies might offer customers competitive non-oil-based alternatives. In order to respond to these constraints, firms might consider their *own alternatives* in terms of investments in technologies that allow fossil fuel switching or efficiency increases. But each of these technologies will be accompanied by different market and price conditions and, thus, management faces uncertainty regarding the *consequences* of individual response options. Most importantly, this also implies that new environmental state and organizational effect uncertainties may need to be addressed. Similarly, the extent of climate change's alteration of the business environment and the resulting economic consequences are still debated (Clarkson and Deyes, 2002; Nordhaus, 2006; Tol, 2003). But how individual firms will be affected by temperature increases is hard to predict. For example, firms can adapt to climate change by relocating production facilities, but it remains uncertain whether such measures will be sufficient to avoid financial disadvantages.

Uncertainties also prevail with respect to institutional change. For example, the *scale and scope of human responses* in the context of the EU ETS are neither stable nor predictable for firms. This pertains to the future development of the EU ETS (e.g., the general amount of available emission allowances), stakeholder pressures on firms to improve their carbon performance and to be in compliance with the EU ETS (e.g., financial markets fostering initiatives such as the Carbon Disclosure Project), as well as the CO_2 allowance price (there is only a limited data basis for anticipating future CO_2 prices).[4] Furthermore, not all firms experience the same *exposure to and relevance* of the institutional change by the EU ETS. For example, for many energy utilities the system provided a positive monetary effect (Sijm et al., 2005); however, it is uncertain whether this will persist in the future. In order to respond to this institutional change, firms could consider carbon capture and storage (CCS) as their *own alternative* to reduce CO_2 emissions. However, the *consequences* are hard to predict especially if underground reservoirs for CO_2 storage leak over time. In addition, it is unclear whether CCS will be acknowledged by the EU ETS and what the consequences of potential leakages on a firm's image would be. Therefore, a CCS investment would be accompanied by new environmental state and organizational effect uncertainties.

Step 3: *assessing the consequences for the determinants of investments' profitability*

In order to determine the profitability of investments in light of carbon constraints, it is important to assess the relevance of each of the above-derived uncertainties and interpret them in terms of probabilities. Depending on their severity, some uncertainties might require a stronger incorporation in terms of future flexible adjustments of the initial investment plan. In analyzing the *underlying investment conditions*, it is important for management to consider all parameters of the project and the value of the investment under current conditions. For instance, uncertainty regarding the general state of oil availability is important for strategic decisions such as whether to invest in oil-based or carbon-free technologies. Uncertainty about the effects of higher oil prices on the firm's future cash flows impedes the accurate analysis of an investment's revenues. Therefore, the likely distribution of future revenue can be determined by the *appraisal of volatilities*. For crude oil, ex post volatilities or those on future markets can be considered. Finally, management has to assign, based on the results of the previous analysis, an optimal *time to invest*.

Step 4: *deriving ecology-driven real options*

We now provide empirical examples where firms considered ecology-driven real options under carbon constraint-related uncertainties. The technology firm Choren Industries considered an *option to defer* regarding an investment in a gasification technology for biomass and carbon-containing residues. The investment was subject to several uncertainties, notably regarding the technology's general feasibility and competitiveness with fuel-based technologies (response uncertainty pertaining to its own alternatives for and consequences of coping with ecology-induced constraints). Nevertheless, the firm patented a specialized gasification process in 1995 (Choren, 2006). With this initial investment, the firm obtained the option to bring this new technology to market once it proved able to compete with traditional processes. Based on the patent, at present Choren builds large-scale industrial plants.

The oil firm Shell incorporated an *option to grow* regarding investments in hydrogen technology. As a response to carbon constraints, the firm assumes that substantial markets for hydrogen-powered vehicles will develop (Shell, 2006). However, these assumptions depend on a range of uncertain factors, one important being the future price of gasoline (state uncertainty pertaining to the extent and timing of ecological limitations). Nevertheless, Shell set up

Shell Hydrogen in 1999 with an initial investment dedicated to investigating business opportunities related to hydrogen and fuel cells as an alternative energy source. By continuously investing in R&D projects and growing Shell Hydrogen, Shell retains the option to participate in the growing hydrogen market at a later stage.

The chemical firm BASF pursued an *option to extend* by investing in its eco-efficiency analysis tool (Saling, 2002). The tool assesses life cycle-wide impacts of products and processes, including the firm's CO_2 emissions and fossil fuel consumption. The motivation behind this initiative can be partly explained as the fulfillment of stakeholder expectations, since the chemical industry is usually considered high polluting. However, it was unclear if stakeholders would consider this approach appropriate and sufficient (decision response regarding their own alternatives for and consequences of adjusting to ecology-induced institutional changes). Nevertheless, BASF started to develop the methodology as an initial investment and later decided to transfer it to other products and processes. Up to 250 analyses have been completed. Furthermore, the method is being offered to other firms through a consultancy service.

The firm Chief Industries embedded an *option to switch* in the case of investing in an ethanol production facility (Nilles, 2006). The facility required a boiler, and management faced the question which fuel would be most cost-effective in the future given emerging carbon constraints (effect uncertainty regarding the magnitude and direction of ecology-induced constraints for the firm). The initial investment was a coal-fired boiler. However, the firm was later able to utilize a dual-fuel system capable of burning coal or natural gas. Depending on current fuel price developments, Chief Industries uses the option to switch between fuels. As a result, this decreases production costs.

The automotive firm Volkswagen utilized an *option to abandon* when it withdrew from the production of the Lupo car (Reed, 2007). Like the automotive industry in general, the firm faces issues in the context of carbon constraints, one of them being consumer preferences about car-specific fuel consumption (effect uncertainty regarding the exposure to and relevance of ecology-induced institutional changes for the firm). The Lupo was produced with the intention of being the world's first car in series consuming as little as 3 l of gasoline/100 km. However, the demand for the car turned out to be rather low and Volkswagen stopped production in 2005. Instead, the firm utilizes the knowledge it acquired to pursue a more successful strategy and introduced the Blue Motion line.

Discussion and conclusions

Within the continuously evolving field of business ethics, managers "confront dilemmas of increasing complexity in a climate of uncertainty and change" (Nicholson, 1994, p. 593). Ecology-induced issues are a prime example for such a change, as they pertain to business conditions that were previously taken for granted. Furthermore, constraints within the natural environment are accompanied by new uncertainties that need to be addressed in a systematic manner. One prominent example is climate change, which is already seen as the most important risk factor for insurance companies (Ernst & Young and Oxford Analytica, 2008). Similarly, institutional action aimed at conserving the natural environment represents another source of new uncertainties that needs to be addressed within corporate risk management. For instance, legislation such as the EU ETS has to be regarded as a critical source of uncertainty with far-reaching strategic implications for firms (compare Hillman and Hitt, 1999). In order to grasp these uncertainties, this article classifies six areas of ecology-induced uncertainties, which challenge strategic management and, notably, corporate investment planning.

For firms to respond to this challenge, we suggest the concept of ecology-driven real options and derive an integrative investment framework. The framework enables the systematic consideration of ecology-induced uncertainties and facilitates the incorporation of flexibility when assessing the profitability of investments. From a competitive advantage point of view, real-options-thinking helps management to re-conceptualize the relevance that issues in the natural environment hold for the business environment. Specifically, it fosters a market-related understanding of emerging uncertainties with respect to issues that currently do not seem market-related and, thus, it supports the self-interest of firms to understand and reduce business risk.

From a business ethics perspective, it contributes to a more sustainable society by enabling management to more quickly and adequately address issues related to the natural environment.

We built our arguments from an ecological economics' point of view and explicitly took an instrumental perspective on firms' responses to a changing natural environment. However, in the context of sustainable development also social issues and voluntary activities of firms influence the corporate business environment (Garriga and Mele, 2004; Husted and Allen, 2007). Voluntary activities in addressing social and environmental issues beyond compliance can be ascribed to CSR (Lo and Sheu, 2007; van Marrewijk and Werre, 2003). However, as with ecology-induced issues, substantial uncertainties emerge, which need to be addressed when governing CSR policies (Lepoutre et al., 2007). One prominent example is the Nike case in which the firm was publicly confronted with child labor practices in its supply chain (De Tienne and Lewis, 2005; Zadek, 2004) and illustrated that strategic management also faces the challenge of ethical decision making (McDevitt et al., 2007). As such, sustainable development with all its different facets increasingly emerges as a business topic (Hart, 1995; Hart and Milstein, 1999; Shrivastava, 1995). Therefore, future research might extend our concept to 'sustainability-driven real options.' However, research has to address notable differences between environmental and social issues (Hannigan, 2006) when analyzing corresponding uncertainties within the business environment.

While we highlight the advantages of adopting a real options logic, the literature also discusses limitations of its practical application. Mostly, real option calculations assume that the underlying asset is similar to a so-called European option, i.e., the option can only be exercised on the expiration day. However, in real life, options are exercised whenever it seems most suitable. Hence, many real options rather resemble American options, which do not require a pre-defined time to invest (Luehrman, 1998), but which are more difficult to calculate. Furthermore, it is also important to be aware that the calculation of real options is based on certain assumptions for probability distributions (Brach, 2003; Copeland and Antikarov, 2001) and for public and private risks (Borison, 2005). Thus, overall it appears to be challenging to empirically calculate all different real options in an exact manner. Nevertheless, due to the increasing salience of ecology-induced issues and the characteristics of the underlying uncertainties, we stress the importance of at least implementing ecology-driven real-options-*thinking* within investment planning. Notably, accelerating climate change and the current consumption rates of fossil fuel urgently need countervailing forces, as emerging carbon constraints constitute a serious issue for society. As such, climate change rightly has to be considered as an ethical issue (e.g., Atkisson, 2007; Le Menestrel et al., 2002). Real-options-thinking can point the way forward, as management might consider investments in low-carbon and low-energy technologies even if these appear not to be profitable under current market conditions.

Acknowledgments

The authors would like to thank Gary R. Weaver, three anonymous JBE reviewers, the participants of the track 'Environmental Issues and Sustainability' at the IFSAM VIIIth World Congress 2006 as well as the participants of the 2008 Research Seminar on 'Business Ethics and Corporate Social Responsibility' at the University of Zurich for their comments and suggestions on previous versions of this paper.

Notes

[1] We use the terms *environment* or *business environment* to refer to the general environment of firms. When referring to the ecological dimension of this environment, we use the term *natural environment*.

[2] It could be argued that these questions are less relevant in times of large-scale economic changes caused by the subprime crisis and the ensuing global financial crisis that have started to unfold in 2008. These changes bear the risk of superseding areas of ethical concern in business decisions. However, due to the increasing ecological challenges that will affect the business environment, it is important to investigate firms' changing comprehension of and motivation to address ecological issues while ascertaining that their behavior can be considered to be "acceptable and appropriate" (Stanwick and Stanwick, 2009, p. 3).

[3] The EU ETS covers over 11,500 energy-intensive installations across the EU and represents almost half of

Europe's CO_2 emissions Installations included are combustion plants, oil refineries, coke ovens, iron and steel plants, and factories making cement, glass, lime, brick, ceramics, pulp, and paper. The first trading period ended in 2007; the second one will last until 2012. For further information, see http://ec.europa.eu/environment/climat/emission.htm.

[4] The theoretical price should reflect the marginal abatement costs (Bailey 1998). However, in reality, CO_2 prices have been determined by a large variety of factors, such as fuel prices, weather conditions, and availability of production capacities (Sijm et al., 2005).

References

Adner, R. and D. A. Levinthal: 2004, 'What is not a Real Option: Considering Boundaries for the Application of Real Options to Business Strategy', *Academy of Management Review* 29(1), 74–85.

Amram, M. and N. Kulatilaka: 1999, *Real Options: Managing Strategic Investment in an Uncertain World* (Harvard Business School Press, Boston).

Aragon-Correa, J. A. and S. Sharma: 2003, 'A Contingent Resource-Based View of Proactive Corporate Environmental Strategy', *Academy of Management Review* 28(1), 71–88.

Atkisson, A.: 2007, 'Global Warming is an Ethical Issue', http://www.earthcharterinaction.org/climate/2007/10/global_warming_is_an_ethical_i.html. Accessed 26 Sept 2008.

Baecker, P. and U. Hommel: 2004, '25 Years Real Option Approach to Investment Valuation: Review and Assessment', in T. Dangl, M. Kopel and W. Kuersten (eds.), *Real Options* (Gabler, Wiesbaden), pp. 1–53.

Bailey, E.: 1998, 'Intertemporal Pricing of Sulfur Dioxide Allowances', Working paper, MIT Center for Energy and Environmental Policy Research Massachusetts.

Bansal, P. and K. Roth 2000, 'Why Companies go Green: A Model of Ecological Responsiveness', *Academy of Management Journal* 43(4), 717–736. doi:10.2307/1556363.

Benford, R. D. and D. A. Snow: 2000, 'Framing Processes and Social Movements: An Overview and Assessment', *Annual Review of Sociology* 26, 611–639. doi:10.1146/annurev.soc.26.1.611.

Black, F. and M. Scholes: 1973, 'Pricing of Options and Corporate Liabilities', *The Journal of Political Economy* 81(3), 637–654. doi:10 1086/260062.

Blyth, W., R. Bradley, D. Bunn, C. Clarke, T. Wilson and M. Yang: 2007, 'Investment Risks Under Uncertain Climate Change Policy', *Energy Policy* 35(11), 5766–5773.

Borison, A. 2005, 'Real Options Analysis: Where are the Emperor's Clothes?', *Journal of Applied Corporate Finance* 17(2), 17–31. doi:10.1111/j.1745-6622.2005.00029.x.

Boyd, B. K., G. G. Dess and A. M. A. Rasheed: 1993, 'Divergence Between Archival and Perceptual Measures of the Environment – Causes and Consequences', *Academy of Management Review* 18(2), 204–226. doi:10.2307/258758.

Brach, M. A.: 2003, *Real Options in Practice* (Wiley, Hoboken).

Buchko, A. A.: 1994, 'Conceptualization and Measurement of Environmental Uncertainty – An Assessment of the Miles and Snow Perceived Environmental Uncertainty Scale', *Academy of Management Journal* 37(2), 410–425. doi:10.2307/256836.

Busch, T. and V. H. Hoffmann: 2007, 'Emerging Carbon Constraints for Corporate Risk Management', *Ecological Economics* 62(3–4), 518–528. doi:10.1016/j.ecolecon.2006.05.022.

Chichilnisky, G. and G. Heal: 1998, 'Global Environmental Risks', in G. Chichilnisky, G. Heal and A. Vercelli (eds.), *Sustainability: Dynamics and Uncertainty* (Kluwer Academic Publishers, Dordrecht).

Choren: 2006, 'The Company Story: Yesterday – Today – Tomorrow', http://www.choren.com. Accessed 7 March 2007.

Clark, W. C.: 1986, 'Sustainable Development of the Biosphere: Themes for a Research Program', in W. C. Clark and R. E. Munn (eds.), *Sustainable Development of the Biosphere* (Cambridge University Press, Cambridge), pp. 1–48.

Clarkson, R. and K. Deyes: 2002, *Estimating the Social Cost of Carbon Emissions* (Department of Environment, Food and Rural Affairs, London).

Clemen, R.: 1996, *Making Hard Decisions – An Introduction to Decision Analysis*, 2nd Edition (Duxbury Press, Belmont).

Common, M. and S. Stagl: 2005, *Ecological Economics: An Introduction* (Cambridge University Press, Cambridge)

Copeland, T. and V. Antikarov: 2001, *Real Options – A Practitioner's Guide* (Texere, New York).

Copeland, T. and P. Keenan: 1998, 'Making Real Options Real', *The McKinsey Quarterly* 3, 128–141.

Cormen, T., C. Leiserson, R. Rivest and C. Stein: 2001, *Introduction to Algorithms*, 2nd Edition (MIT Press, McGraw-Hill, Cambridge, New York).

Cornelius, P., A. Van De Putte and M. Romani: 2005, 'Three Decades of Scenario Planning in Shell', *California Management Review* 48(1), 92–109.

Cortazar, G., E. S. Schwartz and M. Salinas: 1998, 'Evaluating Environmental Investments: A Real Options Approach', *Management Science* 44(8), 1059–1070.

Costanza, R., H. E. Daly and J. A. Bartholomew: 1991, 'Goals, Agenda, and Policy Recommendations for Ecological Economics', in R. Costanza (ed.), *Ecological Economics: The Science and Management of Sustainability* (Columbia University Press, New York), pp. 1–20.

Crane, A. and D. Matten: 2007, *Business Ethics*, 2nd Edition (Oxford University Press, Oxford).

Damanpour, F. and W. M. Evan: 1984, 'Organizational Innovation and Performance – The Problem of Organizational Lag', *Administrative Science Quarterly* **29**(3), 392–409. doi:10.2307/2393031.

Dawkins, J. and S. Lewis: 2003, 'CSR in Stakeholder Expectations: And Their Implication for Company Strategy', *Journal of Business Ethics* **44**(2), 185–193. doi:10.1023/A:1023399732720.

De Tienne, K. B. and L. W. Lewis: 2005, 'The Pragmatic and Ethical Barriers to Corporate Social Responsibility Disclosure: The Nike Case', *Journal of Business Ethics* **60**(4), 359–376. doi:10.1007/s10551-005-0869-x.

DiMaggio, P. J. and W. W. Powell: 1983, 'The Iron Cage Revisited – Institutional Isomorphism and Collective Rationality in Organizational Fields', *American Sociological Review* **48**(2), 147–160. doi:10.2307/2095101.

Dixit, A. K. and R. S. Pindyck: 1994, *Investment Under uncertainty* (Princeton University Press, Princeton).

Dixit, A. K. and R. S. Pindyck: 1995, 'The Options Approach to Capital-Investment', *Harvard Business Review* **73**(3), 105–115.

Donaldson, T. and T. W. Dunfee: 1999, *Ties That Bind: A Social Contracts Approach to Business Ethics* (Harvard Business School Press, Boston).

Ernst & Young and Oxford Analytica: 2008, *Strategic Business Risk: Insurance 2008*. EYGM Limited.

Ferrell, O. C., J. Fraedrich and L. Ferrell: 2002, *Business Ethics. Ethical Decision Making and Cases*, 5th Edition (Hoghton Mifflin, Boston).

Friedman, M.: 1962, *Capitalism and Freedom* (University of Chicago Press, Chicago).

Garrett, T. M.: 1966, *Business Ethics* (Meredith Publishing Company, New York).

Garriga, E. and D. N. Mele: 2004, 'Corporate Social Responsibility Theories: Mapping the Territory', *Journal of Business Ethics* **53**(1–2), 51–71. doi:10.1023/B:BUSI.0000039399.90587.34.

Gladwin, T. N., J. J. Kennelly and T. S. Krause: 1995, 'Shifting Paradigms for Sustainable Development – Implications for Management Theory and Research', *Academy of Management Review* **20**(4), 874–907. doi:10.2307/258959.

Greenwood, R. and C. R. Hinings: 1996, 'Understanding Radical Organizational Change: Bringing Together the Old and the New Institutionalism', *Academy of Management Review* **21**(4), 1022–1054. doi:10.2307/259163.

Greenwood, R., R. Suddaby and C. R. Hinings: 2002, 'Theorizing Change: The Role of Professional Associations in the Transformation of Institutionalized Fields', *Academy of Management Journal* **45**(1), 58–80. doi:10.2307/3069285.

Hambrick, D. C., S. Finkelstein and A. C. Mooney: 2005, 'Executive Job Demands: New Insights for Explaining Strategic Decisions and Leader Behaviors', *Academy of Management Review* **30**(3), 472–491.

Hannigan, J.: 2006, *Environmental Sociology*, 2nd Edition (Routledge, New York).

Hart, S. L.: 1995, 'A Natural-Resource-Based View of the Firm', *Academy of Management Review* **20**(4), 986–1014. doi:10.2307/258963.

Hart, S. L. and M. B. Milstein: 1999, 'Global Sustainability and the Creative Destruction of Industries', *Sloan Management Review* **41**(1), 23–33.

Heal, G.: 1998, 'Interpreting Sustainability', in G. Chichilnisky, G. Heal and A. Vercelli (eds.), *Sustainability: Dynamics and Uncertainty* (Kluwer Academic Publishers, Dordrecht).

Hillman, A. J. and M. A. Hitt: 1999, 'Corporate Political Strategy Formulation: A Model of Approach, Participation, and Strategy Decisions', *Academy of Management Review* **24**(4), 825–842. doi:10.2307/259357.

Hirsch, R., R. Bezdek and R. Wendling: 2005, *Peaking of World Oil Production: Impacts, Mitigation, and Risk Management*. DOE NETL.

Hoffman, A. J.: 1999, 'Institutional Evolution and Change: Environmentalism and the US Chemical Industry', *Academy of Management Journal* **42**(4), 351–371. doi:10.2307/257008.

Hourcade, J. C., D. Demailly, K. Neuhoff and M. Sato: 2007, 'Differentiation and Dynamics of EU ETS Industrial Competitiveness Impacts', Climate Strategies Report.

Husted, B. W.: 2005, 'Risk Management, Real Options, and Corporate Social Responsibility', *Journal of Business Ethics* **60**(2), 175–183. doi:10.1007/s10551-005-3777-1.

Husted, B. W. and D. B. Allen: 2007, 'Corporate Social Strategy in Multinational Enterprises: Antecedents and Value Creation', *Journal of Business Ethics* **74**(4), 345–361. doi:10.1007/s10551-007-9511-4.

IPCC: 2007, *Climate Change 2007: The Physical Science Basis* (Cambridge University Press, Cambridge).

Jones, T. M.: 1991, 'Ethical Decision-Making by Individuals in Organizations – An Issue-Contingent Model', *Academy of Management Review* **16**(2), 366–395. doi:10.2307/258867.

King, A.: 1995, 'Avoiding Ecological Surprise – Lessons from Long-Standing Communities', *Academy of Management Review* **20**(4), 961–985. doi:10.2307/258962.

Knouse, S. B. and R. A. Giacalone: 1992, 'Ethical Decision-Making in Business – Behavioral Issues and Concerns', *Journal of Business Ethics* **11**(5–6), 369–377. doi:10.1007/BF00870549.

Kuik, O.: 2003, 'Climate Change Policies, Energy Security and Carbon Dependency', *International Environmental Agreements: Politics, Law and Economics* **3**, 221–242. doi:10.1023/B:INEA.0000005625.44125.54.

Laurikka, H. and T. Koljonen: 2006, 'Emissions Trading and Investment Decisions in the Power Sector—A Case Study in Finland' *Energy Policy* **34**(9), 1063–1074.

Lawrence, A. T., J. Weber and J. E. Post: 2005, *Business and Society: Stakeholder Relations, Ethics, Public Policy*, 11th Edition (McGraw-Hill, New York).

Le Menestrel, M., S. van den Hove and H. C. de Bettignies: 2002, 'Processes and Consequences in Business Ethical Dilemmas: The Oil Industry and Climate Change', *Journal of Business Ethics* **41**(3), 251–266. doi:10.1023/A:1021237629958.

Lepoutre, J., N. A. Dentchev and A. Heene: 2007, 'Dealing with Uncertainties When Governing CSR Policies', *Journal of Business Ethics* **73**(4), 391–408. doi:10.1007/s10551-006-9214-2.

Lin, T. T., C. C. Ko and H. N. Yeh: 2007, 'Applying Real Options in Investment Decisions Relating to Environmental Pollution', *Energy Policy* **35**(4), 2426–2432.

Lo, S. F. and H. J. Sheu: 2007, 'Is Corporate Sustainability a Value-Increasing Strategy for Business?', *Corporate Governance: An International Review* **15**(2), 345–358. doi:10.1111/j.1467-8683.2007.00565.x.

Luehrman, T. A.: 1998, 'Investment Opportunities as Real Options: Getting Started on the Numbers', *Harvard Business Review* **76**(4), 51–67.

McDevitt, R., C. Giapponi and C. Tromley: 2007, 'A Model of Ethical Decision Making: The Integration of Process and Content', *Journal of Business Ethics* **73**(2), 219–229. doi:10.1007/s10551-006-9202-6.

Meadows, D. H., D. L. Meadows, J. Randers and W. W. Behrens: 1972, *The Limits to Growth* (Universe Books, New York).

Miller, D. and J. Shamsie 1999, 'Strategic Responses to Three Kinds of Uncertainty: Product Line Simplicity at the Hollywood Film Studios', *Journal of Management* **25**(1), 97–116. doi:10.1177/014920639902500105.

Milliken, F. J.: 1987, '3 Types of Perceived Uncertainty About the Environment – State, Effect, and Response Uncertainty', *Academy of Management Review* **12**(1), 133–143. doi:10.2307/257999.

Nicholson, N.: 1994, 'Ethics in Organizations – A Framework for Theory and Research', *Journal of Business Ethics* **13**(8), 581–596. doi:10.1007/BF00871806.

Nilles, D.: 2006, 'Process Heat and Steam Alternatives Rising', *Ethanol Producer Magazine*, http://www.ethanolproducer.com. Accessed 30 March 2007.

Nordhaus, W. D.: 2006, 'The Stern Review on the Economics of Climate Change', NBER Working Paper No. W12741, National Bureau of Economic Research, Cambridge.

Oliver, C.: 1992, 'The Antecedents of Deinstitutionalization', *Organization Studies* **13**(4), 563–588. doi:10.1177/017084069201300403.

Oliver, C.: 1997, 'Sustainable Competitive Advantage: Combining Institutional and Resource-Based Views', *Strategic Management Journal* **18**(9), 697–713. doi:10.1002/(SICI)1097-0266(199710)18:9<697::AID-SMJ909>3.0.CO;2-C.

Point Carbon: 2008, 'Data', *Trading Carbon* **02**(03), 38–39.

Polletta, F. and J. M. Jasper: 2001, 'Collective Identity and Social Movements', *Annual Review of Sociology* **27**, 283–305. doi:10.1146/annurev.soc.27.1.283.

Porter, M. E. and C. van der Linde: 1995, 'Green and Competitive – Ending the Stalemate', *Harvard Business Review* **73**(5), 120–134.

Prigogine, I. and I. Stengers: 1984, *Order out of Chaos: Man's New Dialogue with Nature* (Bantam Books, Toronto).

Reed, J.: 2007, 'Problems of Pitching Cleaner Cars to the Unconverted', *Financial Times (North American Edition)* Jan, 30 (London).

Rugman, A. M. and A. Verbeke: 1998, 'Corporate Strategies and Environmental Regulations: An Organizing Framework', *Strategic Management Journal* **19**(4), 363–375. doi:10.1002/(SICI)1097-0266(199804)19:4<363::AID-SMJ974>3.0.CO;2-H.

Sadorsky, P.: 1999, 'Oil Price Shocks and Stock Market Activity', *Energy Economics* **21**(5), 449–469. doi:10.1016/S0140-9883(99)00020-1.

Saling, P.: 2002, 'Life Cycle Management, Eco-efficiency Analysis by BASF: The Method', http://corporate.basf.com. Accessed 30 March 2007.

Schwartz, P.: 2007, 'Investing in Global Security', *Harvard Business Review* **85**(10), 26–27.

Scott, W. R.: 2001, *Institutions and Organizations*, 2nd Edition (Sage Publications, London).

Sharp, F. C. and P. G. Fox: 1937, *Business Ethics. Studies in Fair Competition* (D. Appleton-Century Company, New York).

Shell: 2006, 'Shell Hydrogen', http://www.shell.com. Accessed 25 Oct 2006

Shrivastava, P.: 1995, 'Environmental Technologies and Competitive Advantage', *Strategic Management Journal* **16**, 183–200. doi:10.1002/smj.4250160923.

Sijm, J. P. M., S. J. A. Bakker, Y. Chen, H. W. Harmsen and W. Lise: 2005, CO_2 *Price Dynamics: The Implications* of EU Emissions Trading for the Price of Electricity (Energy Research Centre of the Netherlands, Petten).

Smircich, L. and C. Stubbart: 1985, 'Strategic Management in an Enacted World', *Academy of Management Review* **10**(4), 724–736. doi:10.2307/258041.

Stanwick, P. A. and S. D. Stanwick: 2009, *Understanding Business Ethics* (Pearson Prentice Hall, Upper Saddle River, NJ).

Starik, M.: 1995, 'Should Trees Have Managerial Standing – Toward Stakeholder Status for Nonhuman Nature', *Journal of Business Ethics* **14**(3), 207–217. doi:10.1007/BF00881435.

Stern, N.: 2006, *The Economics of Climate Change – The Stern Review* (Cambridge University Press, Cambridge).

Thompson, J. D.: 1967, *Organizations in Action* (McGraw-Hill, New York).

Tol, R. S. J.: 2003, 'Is the Uncertainty About Climate Change too Large for Expected Cost-Benefit Analysis?', *Climatic Change* **56**(3), 265–289. doi:10.1023/A:1021753906949.

Trevino, L. K.: 1986, 'Ethical Decision-Making in Organizations – A Person-Situation Interactionist Model', *Academy of Management Review* **11**(3), 601–617. doi:10.2307/258313.

Trevino, L. K. and S. A. Youngblood: 1990, 'Bad Apples in Bad Barrels – A Causal-Analysis of Ethical Decision-Making Behavior', *The Journal of Applied Psychology* **75**(4), 378–385. doi:10.1037/0021-9010.75.4.378.

Trigeorgis, L.: 1988, 'A Conceptual Options Framework for Capital Budgeting', *Advances in Futures and Options Research* **3**, 145–167.

Trigeorgis, L.: 1995, *Real Options in Capital Investment: Models, Strategies, and Applications* (Praeger, Westport & London).

van Marrewijk, M. and M. Werre: 2003, 'Multiple Levels of Corporate Sustainability', *Journal of Business Ethics* **44**(2), 107–119. doi:10.1023/A:1023383229086.

Wheatley, M. J.: 1999, *Leadership and the New Science: Discovering Order in a Chaotic World*, Vol. 2 (Berrett-Koehler Publishers, San Francisco).

Winn, M., and M. Kirchgeorg: 2005, 'The Siesta is over: A Rude Awakening from Sustainability Myopia', in S. Sharma and M. Starik (eds.), *Research in Corporate Sustainability, Volume 3, Strategic Capabilities and Competitiveness* (Elgar, Northampton), pp. 232-258.

Yang, M., W. Blyth, R. Bradley, D. Bunn, C. Clarke and T. Wilson: 2008, 'Evaluating the Power Investment Options with Uncertainty in Climate Policy', *Energy Economics* **30**(4), 1933–1950.

Zadek, S.: 2004, 'The Path to Corporate Responsibility', *Harvard Business Review* **82**(12), 125–132.

ETH Zurich,
Zurich, Switzerland
E-mail: tobusch@ethz.ch

i want morebooks!

Buy your books fast and straightforward online - at one of world's fastest growing online book stores! Environmentally sound due to Print-on-Demand technologies.

Buy your books online at
www.get-morebooks.com

Kaufen Sie Ihre Bücher schnell und unkompliziert online – auf einer der am schnellsten wachsenden Buchhandelsplattformen weltweit! Dank Print-On-Demand umwelt- und ressourcenschonend produziert.

Bücher schneller online kaufen
www.morebooks.de

VDM Verlagsservicegesellschaft mbH
Heinrich-Böcking-Str. 6-8   Telefon: +49 681 3720 174   info@vdm-vsg.de
D - 66121 Saarbrücken   Telefax: +49 681 3720 1749   www.vdm-vsg.de

Printed by Books on Demand GmbH, Norderstedt / Germany